# History of the Electric Automobile
## Hybrid Electric Vehicles

Other SAE books of interest on this topic:

**Alternative Cars in the 21ˢᵗ Century**
Robert Q. Riley
(Order No. R-139)

**Building the E-motive Industry**
**Essays and Conversations About Strategies for Creating an Electric Vehicle Industry**
Scott A. Cronk
(Order No. R-148)

**Electric Vehicles**
**Driving Towards Commercialization**
Edited by Ron Sims and Bradford Bates
(Order No. PT-58)

For information on these or other related books, contact:
Society of Automotive Engineers Inc.
400 Commonwealth Drive
Warrendale, PA 15096-0001 U.S.A.
Phone: (724) 776-4970
Fax: (724) 776-0790
E-mail: publications@sae.org

# History of the Electric Automobile
## Hybrid Electric Vehicles

Ernest Henry Wakefield, Ph.D.

Society of Automotive Engineers, Inc.
Warrendale, Pa.

**Library of Congress Cataloging-in-Publication Data**

Wakefield, Ernest Henry, 1915-
    History of the electric automobile : hybrid electric vehicles / Ernest Henry
Wakefield.
      p. cm.
    Includes bibliographical references and index.
    ISBN 0-7680-0125-0
    1. Automobiles, Electric--History. 2. Hybrid Electric Cars--History.  I. Title.
TL220.W343  1988
629.22'93--DC21
                                                    98-3420
                                                      CIP

Vehicles pictured on cover (clockwise from upper left): Esoro H301 Family hybrid electric car; General Motors 1987 Sunraycer; American fuel cell bus; Woods' dual-powered 14-hp coupé of 1917.

Society of Automotive Engineers
400 Commonwealth Drive
Warrendale, PA 15096-0001  U.S.A.
Phone: (724) 776-4841
Fax: (724) 776-5760
E-mail: publications@sae.org
http://www.sae.org


SAE Order No. R-187

*Dr. Ferdinand Porsche (1875–1952), although not the first with hybrid electric vehicles, probably carried the concept furthest of early pioneers at the turn of the century and later.* Automobile Design, *Society of Automotive Engineers, Warrendale, PA, 1992, p. 25.*
*(Courtesy: Klaus Parr, Porsche)*

To

**Unknown**
who circa 5000 years ago
conceived the flywheel as an energy storage device

**John Barber**
who in 1791 was granted a patent for a gas turbine

**Count Alessandro Volta**
who in 1800
discovered the electric battery

**Robert Stirling**
who in 1816
invented the Stirling engine

**Michael Faraday**
who in 1821
first obtained mechanical rotation from an electromagnet assembly

**Alexandre-Edmond Becquerel**
who in 1839
first noticed the photoelectric effect

**Sir William Grove**
who in 1839
assembled the first fuel cell

**Etienne Lénoir**
who circa 1860
built the first internal combustion engine

**Count Felix Carli**
who in 1894
first assembled a hybrid electric car

**All made possible the hybrid electric automobile**

To write it, it took three months (ten years);
to conceive it—three minutes;
to collect all the data in it—all my life.

—F. Scott Fitzgerald

# Table of Contents

# Foreword

For almost four decades, Dr. Ernest Wakefield has been connected with electric vehicles in one way or another—from making them to studying their origins. Although he has been an electrical engineer, he now is pursuing a writing career. His written work covers a range of topics and styles, including fiction. Since 1977, he has been developing books on electrical automobiles. In 1994, Dr. Wakefield published a book on the history of electric automobiles, which was devoted to battery-only power. The present book is also a historical account of electric automobiles and covers hybrid propulsion, involving electricity, as well as mechanical devices such as flywheels, the new gasoline engine-battery vehicles, and solar vehicles. Many readers will enjoy the discussions of the various solar car competitions. Likewise, do not miss the appendices, which have wonderful additional information.

The 1970 Act of Congress mandating research in this area has certainly propelled this topic into clear view here in the United States. Therefore this book will be of widespread interest. It is a useful combination of the history and technical aspects of electric vehicles, including many marvelous photographs, sketches, and technical information.

Much of this work has been completed here in the halls of "Tech," the building occupied by the McCormick School of Engineering and Applied Science—Northwestern University while Dr. Wakefield has been a quiet guest. As a midwestern engineering school, we have a traditional and special love of things automotive, and we are proud to have been of some help in producing this work. We are pleased to have Dr. Wakefield's expertise and knowledge with us. We are all well served when an engineer such as Dr. Wakefield can communicate so well and uses his research skills to investigate this theme, presenting it to us with technical clarity and historical insight.

*Jerome B. Cohen, Dean*
McCormick School of Engineering and Applied Sciences
Northwestern University

# Preface

To many, the romance and nostalgia of electric cars and electric carriages have an almost mystical interest. Moreover, with electricity being such a clean and versatile fuel that is widely used, a perennial question is: why not employ it for personal transportation? For more than 30 years, I have been an integral part of the emerging electric vehicle industry in conception, design, construction, marketing, financing, and chronicling its development.

The authorship of *History of the Electric Automobile*: *Hybrid Electric Vehicles* has been a labor of love that is hoped to be objective and correct. As you read this book or the previous volume, *History of the Electric Automobile: Battery-Only Powered Cars*, it becomes apparent that electric cars result from a happy arrangement of mechanical, electrical, and chemical laws which, operating in perfect harmony, yield personal transportation. The earlier volume describes the historical development of chargers, batteries, controllers, and motors, as well as the complete battery-only powered car. However, the *History of the Electric Automobile*: *Hybrid Electric Vehicles* confines itself to the application of capacitors, engines (Otto, Stirling, and turbine), flywheels, fuel cells, inductive charging, and solar cells to hybrid electric vehicles. It also outlines in detail their performance, mentions those investigators responsible for them, and hints on their charging, design, and construction.

This book may be divided into four segments. First is a brief introduction to indicate how the necessary knowledge for electric vehicles was gained through earlier centuries, particularly the period beginning with quantitative experimentation utilizing electrical phenomena. These innovations were finally assembled as an electric vehicle in 1881. Electric automobile developments in both America and Europe are seen when the three sources of energy were in contention: steam, electricity, and gasoline. In design, the automobiles of this period were wagons or carriages closely resembling the horse-drawn vehicles being supplanted. With the discovery of automobile touring in 1895 by Hiram Percy Maxim, the shift toward gasoline-fueled vehicles, capable of long-distance travel, led to their growing dominance. That time was circa 1902. As early as 1894, Count Felix Carli of Italy recognized that the Achilles' heel of the electric car was its limited range—and he did something about it, as described in Chapter 5. Others also tried to overcome this limitation, even to the present time. The history of this substantial effort has led to the *History of the Electric Automobile: Hybrid Electric Vehicles*.

To overcome this limited-range pariah, the second section of this book discusses the potential energy of a wound spring (Carli), the use of gasoline or natural gas engine-electric systems, the Stirling engine-electric approach, the rotational energy of a flywheel, the battery-battery concept, the fuel cell-battery application, the application of inductively charged enhancements, and finally the application of solar cells as range extenders for battery equipped-electric cars.

The third section of the book investigates how pollution from the ubiquitous gasoline-engined car, together with recent discoveries in enhanced efficiencies from the photoelectric effect, have spawned interest in racing of battery-carrying, solar-cell-equipped electric vehicles. Outlined here are experiences being gained in these solar-assist electric-powered vehicles and the possibility of their commercialization.

The Appendices, the fourth section of the book, describe the Hughes inductive charging system and the employment of the capacitor with regenerative braking.

From 1881 to the present, essentially all seers predicting the role of electric vehicles in personal transportation have erred. Therefore, *History of the Electric Automobile: Hybrid Electric Vehicles* attempts to record the significant developments in many types of road-using, hybrid electric vehicles, both domestic and international which, if not captured in a book, may indeed be lost.

# Acknowledgments

Few books are written without help and guidance from others. In preparing *History of the Electric Automobile: Hybrid Electric Vehicles*, I am indebted to many in Australia, England, France, Germany, Italy, Japan, the Netherlands, Sweden, Switzerland, and the United States, as cited below. Names of unlisted persons may possibly be found with the figures throughout the book. Surely some contributors have been inadvertently omitted. We know a mere acknowledgment is too little acclaim for their interests.

Listed alphabetically by nations, those in Australia are: Editor, *The Advertiser* (Adelaide, Australia), for providing newspaper copy of the 1987 Trans-Australian Race; Christine Burnup, General Manager, Public Relations, Pasminco, Ltd., Melbourne; Peter O. Fries of St. Lucia, Queensland, author of *The Technical Report (on the 1987 Trans-Australian Solarmobile Race) for the New Energy Development Organization of Japan*; Jon D. Retford, Project Engineer for Ford Australia's *Sunchaser* solarmobile, for providing technical information; A.E. Ryan, Public Affairs Office of Ford Motor Company of Australia, for kindly furnishing data and illustrations of its *Sunchaser* solarmobile; Christopher Tayle, Editor, *Darwin* (Australia) *Northern Territorian,* for providing copy on the 1987 Trans-Australian Race*;* David Taylor, Syndication Manager, *The West Australian,* Perth, for photos; Hans Tholstrup, who promoted the (Australian) Pentax World Solar Challenge Race and who deserves lasting credit; and Bill Tuckey, who produced the elegant book, *Sunraycer's Solar Saga,* and generously provided pictures.

Those from England are: Timothy Auger, Editorial Director of B.T. Batsford Ltd., publishers of *Racing and Sports Car Chassis Design,* who aided me in locating Michael Costins and David Phipps, the authors. They generously allowed me to use sections of their volume. Thanks are extended to Don Gribble, editor of *Batteries International*, Surrey, England, who allowed me to quote Dr. Victor Wouk's article on the Hughes Magne-Charge. Also from Albion, John Parry of Parry People Movers, Ltd., should be recognized for providing data and photos on his company's flywheel tram.

Assistance from France was provided by Professor J.C. Marpinard of the *Centre National de la Recherche Scientifique Labatoire d'Analyse et d'Architecture des Systèmes*, Toulouse, for a drawing of the petro-electric car.

Those from Germany are: Dr. S. Brüdgman and V. Platzer of Volkswagen for illustrations on the *Chico*, and Helmut Werner of Mercedes for data on Swatch cars. The Public Relations Department of Adam Opel AG and Paul Achihhofer of Ford-Werke AG have been helpful.

Those from Japan are: Masaru Otaki of the Tokyo Eizosha Co., Ltd., Tokyo, for permission to use excerpts from *The Technical Report for the New Energy Development Organization of Japan*; and Masashi Aihara of Mazda (North America) Inc.

Those from the Netherlands are: Frans J.M. Thoolen from Horn, for information and illustrations on flywheels taken from his doctoral thesis; and the Press Department of the Philips Company, Eindhoven, for the picture of Dr. Robert Stirling shown in Chapter 8.

Individuals from Switzerland are: Herzlich Ihr, Editor, *Schweizer* (Swiss) *Illustrierte,* likewise for making available copy and illustrations of the 1987 Swiss *Tour De Sol*; Urs Muntwyler, Manager, Swiss *Tour de Sol*, who deserves credit for promoting solar transportation; and Dr. Fredy Sidler, Director, and Markus Liniger of Ingenieurschule Biel, for photos and engineering data.

Those from the United States are: Charlene Alfonso of Unique Mobility, Inc., for providing photos; Robert R. Aronson, President of Electric Auto Corporation, Fort Lauderdale, Florida, for furnishing prints; Bradford Bates, Manager, Research Laboratory, Alternative Power Sources Technology, Ford; Robert Bockner of Solar Systems Inc., Dodgeville, Wisconsin, for providing information on early solarmobiles; Arthur Boyt of Crowder College in Neosho, Missouri, for writing to me about his experiences with solarmobiles; Dr. Alec N. Brooks, *Sunraycer* Project Manager, AeroVironment, Inc., Monrovia, California, for furnishing technical information and for graciously and critically reading a large section of the manuscript and offering many helpful suggestions; Curt Brown, Eagle-Picher Industries, Inc., of Joplin, Missouri, for supplying technical information on silver-zinc and silver-cadmium secondary cells; Dr. Paul J. Brown, U.S. Department of Energy, Washington, DC, for conversations and for providing reports on electric vehicles; Robert Thomas Chojnacki, Programmer/Analyzer at the Center for Manufacturing Engineering, Northwestern University, for discussions on computer-aided design (CAD) and computer-aided engineering (CAE); Professor K. Ehmann, Northwestern University, for consultation; Ramon Dominguez, Vice President and General Manager of Solarex of Rockwell, Maryland, for explaining the manufacturing operations of making silicon cells for satellite and solarmobile use; and the late Claud Erickson, then Manager of the Lansing, Michigan, Electric Power Board, for originally interesting me in the history of electric vehicles.

Also, I wish to thank Peter Franco of Evanston, Illinois, for shooting many photos; Douglas A. Fraser, Project Coordinator, Thayer School of Engineering, Dartmouth College, for demonstrating its *Sunvox-I* solarmobile and advising of his subsequent *Sunvox II*, for providing drive-system electrical schematics and text, for discussing solarmobile racing principles, and for kindly reading carefully much of the manuscript; John V. Goldsmith, Vice President of the

Solarex Corporation, for providing permission to generously excerpt its book, *The Solarex Guide to Solar Electricity,* to form parts of Chapters 10 and 11; Terell W. Gunter, Vice President of Indianapolis Motor Speedway Corporation; Julio E. Hamel, then General Manager of the University Consortium of Structural Dynamics Research Corporation of Milford, Ohio; John J. Hanley, CEO of *Scientific American,* for allowing use of illustration of solar cells; Professor Richard S. Hartenberg, Northwestern University, for supplying illustrations for Chapters 3 and 4 and for keeping me informed; Nancy Hazard, Co-Director of the Northeast Sustainable Energy Association, for sections of Chapter 14; Jan F. Herbst of the Physics Department of General Motors Corporation, Anderson, Indiana, for discussing constituencies of rotors for synchronous motors; Kalwey A. Johnson, Director of Operations of Solarex, for showing me the entire integrated operation for manufacturing silicon cell arrays suitable for solarmobiles and other uses; Dr. Robert S. Kirk of the U.S. Department of Energy, for photos; Professor Alan L. Kistler, Northwestern University, for contributing a section on aerodynamics, providing many discussions, and lending appropriate books; Dr. Philip T. Krein, Professor of Electrical and Computer Science, University of Illinois, Urbana-Champaign, for discussing its petro-electric car; Professor Elmer L. Lewis, Mechanical Engineering Department, Northwestern University, for discussions on flywheels; J. Bruce McCristal, Director of Public Affairs, GM Hughes Electronics, Inc., Detroit, for providing information on the General Motors *Sunraycer*, for furnishing many photos of participating solarmobiles in the 1987 Trans-Australian Race, and for kindly presenting me with a copy of *Sunraycer's Solar Saga;* Robert S. McKee, President, McKee Engineering Company, Lake Zurich, Illinois, for providing photos; Dr. Roelf J. Meijer, Founder, Stirling Thermal Motors, Inc., of Ann Arbor, Michigan, for contributing to Chapter 8 on the Stirling engine; and James F. Miller of Argonne National Laboratory, for making a contribution on lithium-based batteries.

Dr. Gordon J. Murphy, who is Professor of Electrical Engineering and Computer Science at Northwestern University has been especially helpful. I spent many years in close association with Dr. Murphy designing and crafting electric vehicles. Dr. Paul A. Nelson, Deputy-Director of the Electrochemical Division, Argonne National Laboratory, provided data on advanced batteries. Furquab Nazeeri of the University of Michigan also provided data. Thanks are extended to Professor Panos Y. Papalambros, University of Michigan, for furnishing some text; P.B. Patil of the Ford Research Laboratory, Dearborn, Michigan, for making available information on sodium-sulfur batteries; Rick Paul of *Motor Trend*, for kindly furnishing information on the Honda *Dream;* Sally Pobojewski, Senior Science Writer of *Michigan Today* of the University of Michigan; Martin Paul, Electronic Publishing, *Scientific American*; Bonne W. Posma, President of Saminco Inc., Fort Meyers, Florida, for hydrogen fuel cell bus photos and others; Dr. Peter Roll, then Vice President for Information, Northwestern University, for designating me an Independent Scholar, thus making computer equipment available to me at the university; and Professor Gene Smith of the University of Michigan, for kindly reading *History of the Electric Automobile: Battery-Only Powered Cars.* Robert P. Sokol of Northwestern University was helpful for automobile information, and Professor R. Rhoads Stephenson, California Institute of Technology, was particularly generous for allowing me to quote sections of his summary on heat engines.

Michael Valenti, Associate Editor, *Mechanical Engineering,* kindly permitted me to quote from his article on the Volvo Environmental Concept Car. I particularly wish to thank Professor Harry West of the Massachusetts Institute of Technology for providing his report, including photos on the 1987 Swiss *Tour de Sol* race. Kenneth Wipke, National Renewable Energy Laboratory in Golden, Colorado, kindly furnished information on the 1993 Ford Electric Hybrid Challenge. Catherine W. Wolf of the Dartmouth College News Service furnished information and pictures about its solarmobile. James D. Worden, then (1988) a student at Massachusetts Institute of Technology, in his pioneering work, has directed my attention to many niceties in the design of solarmobiles. I particularly wish to thank Dr. Victor Wouk of Victor Wouk Associates for reviewing the entire text.

I also owe thanks to the following from Northwestern University for researching the field: Janet Ayres, Engineering and Science Librarian (retired); Janice Dluzynski, Head of the Transportation Library; Connie Avildsen, Jordan Ellison, Caryn Olczyk, Mary McCreadie, Renée McHenry, Hema Ramachadran, Dorothy Ramm, and Mary Roy (retired), Librarians in the Transportation Library; Marjorie Carpenter, then Interlibrary Loan Librarian, and Maude M. Kelso, Periodical Supervisor (retired). Also, I wish to thank the staff of the University of Illinois Library, Urbana-Champaign, for many courtesies.

In reproducing this manuscript, most helpful were the services of Professor Kornel F. Ehmann, Mechanical Engineering Department; Mildred L. Wiesser; Robert T. Schreiber of the Center for Manufacturing Engineering; and Frances Glass-Newman and Elizabeth A. Martin of the Program of Master of Management in Manufacturing, all of Northwestern University in Evanston, Illinois. Robert C. Michaelson, Library Department Head/UL Science Library Head, and the staff of the Northwestern University Engineering Library have been helpful, and their source of reference books has been appreciated. Claudia Wintergerst, of Apple Computer, Inc., and the Mechanical Engineering Department of Northwestern University provided computer equipment. Joel D. Meyer, Assistant Dean, Robert R. McCormick School of Engineering and Applied Sciences, Northwestern University, and the many members of the Information Center Laboratory, Northwestern University, provided answers to questions on computers and allowed me to use their excellent facilities. Finally, I would like to thank Jerome B. Cohen, Dean of the McCormick School of Engineering and Applied Sciences, for providing space and permission to use the knowledge of his faculty to answer numerous technical questions.

Without the above contributors and references, this book would possess a less rich background. Any errors, of course, are mine.

# CHAPTER 1

# Background for the Hybrid Electric Horseless Carriage

Civilization was awaiting the motorcar, and its rapid adoption indicated its superiority over the horse-drawn carriage. As is often true during immaturity of an invention, there are competing systems which, for personal travel at the turn of the century, were carriages powered by steam, by gasoline, and by electricity. This book relates to the hybrid electric vehicles and their mentors. Presented here is background for the newly emerging interest in the electric car as an alternate source of personal transportation. From this and my related volumes,[1,2] you will see that personal electric transit is a logical outgrowth of understanding nature's laws, and it has a long and interesting history.

## Establishing the Principles of Electricity

Little did the ancients realize their knowledge of electric fish, amber, or lodestones would usher in the Electric Age. They knew that amber (the Greek word is *elektron)*, if rubbed, would attract bits of straw. They also observed that a lodestone, a natural magnet, had peculiar properties. With this background and an inquiring mind, a German physicist named Otto von Guericke developed the first static electric machine.[3] On this machine, Hugo von Kleist learned he could charge a glass bottle.[4] Because the experiments were executed at the University of Leyden in Holland, this early electric capacitor was called a Leyden jar. Benjamin Franklin would take this newly devised and useful container, charge it from a string and key connected to a kite flying in a lightning storm, and thus relate electricity and lightning while simultaneously

1

determining the existence of what Franklin elected to call positive and negative charges.[5] With this background and Luigi Galvani's observation that a frog's leg twitched if probed by two different metals,[6] Count Alessandro Volta developed the electric battery capable of yielding a continuous flow of current.[7]

Using this latter reliable source of electrical energy, Michael Faraday opened a new area of natural phenomena. Faraday's inquisitive mind, operating in the era of enlightenment of the early nineteenth century, observed that if a wire, closed on itself through a measuring instrument, were to pass the ends of a magnet, current would flow in the wire. Conversely, if a wire carrying a current were placed in a magnetic field, a force acted on the wire. From these observations, Faraday established the principles of the direct-current motor and obtained rotational motion.[8] Faraday also discovered the principle that a current changing within a wire would induce a voltage in a nearby circuit. This observation had to wait 55 years, until 1888, for the Croatian Nikola Tesla to develop with George Westinghouse the modern alternating-current induction motor, which now powers the factories of the world[9] and currently an increasing number of electric automobiles.

In France, a youthful André Ampère, reading of Faraday's work, investigated the forces interacting between two wires carrying electric charge, and he developed quantitative means of measuring current flow.[10] As if to demonstrate the universality of science, the German Georg Ohm, by repeated experiments with direct current, verified a useful relationship among voltage, current, and resistance of a wire. These observations resulted in Ohm's law

$$E = IR$$

where E, I, and R are voltage, current, and resistance, respectively. Utilizing this simple relationship, if any two quantities are known, the third can be calculated.[11]

Just as this relationship would provide the electric automotive designers a necessary tool, in the same year of Ohm's work (1827), a Frenchman named Onesiphore Pecqueur invented the differential gearing almost universally used with four-wheeled vehicles.[12] Continuing the onward march of science, the English mathematician James Clerk Maxwell elegantly correlated the concept of electric and magnetic fields that extend into space.[13] Building on this knowledge, the German Heinrich Hertz proved the electromagnetic nature of light.[14] He established by experiment what Maxwell had shown to be true mathematically—that electromagnetic radiation could be propagated through space.

These experiments led directly to Guglielmo Marconi making the first two-way commercial radio transmission between Europe and the United States on 18 January 1902.[15] One of the most spectacular applications of Marconi's invention, ship-to-shore transmission, occurred on 14 April 1912 when the *Titanic* collided with an iceberg and 1,507 lives were lost because of an insufficient number of lifeboats. Within minutes after the collision, many ships were racing to the stricken liner.

As essential as was the theory and practice of electrical phenomena, for propulsion there must be an instrument to propel. The first successful experimenters with electric vehicles were Gustave Trouvé, an electrical engineer from France, in 1881; two English professors, William Ayrton and John Perry, in 1882; and Andrew Riker, an American, in 1890. All used the tricycle, a derivative of the bicycle. The origin of the bicycle, a remarkable machine, is given as follows.

# Development of the Bicycle and the Tricycle

While the steam road carriage was first developed without the bicycle or tricycle as a model, both the electric and gasoline-powered vehicles came sufficiently late in the nineteenth century to benefit from these two remarkable types of personal transportation. Of the two, the bicycle came before the three-wheeler. As finally developed, the bicycle is a most efficient machine. On a good road, compared with walking, a person bicycling consumes only one-fifth the energy in traversing an equal distance and at five times the speed.[16] A bicycle will carry ten times its own weight, a factor probably unique in any type of transportation even today.[17] Compared with preparing a horse for riding or drawing a buggy, the time required to place a bicycle or tricycle in operation was smartly reduced. Moreover, lightweight bicycles became relatively inexpensive.

The antecedent of the bicycle was called a *Draisienne,* or *Hobby Horse.* As shown in *History of the Electric Automobile: Battery-Only Powered Cars,*[2] it consisted of two wheels interconnected by a crossbar on which a seat was mounted. The rider propelled both the vehicle and himself by foot action on the ground. This vehicle was invented by Nicephore Niepce of Chalons, France, in 1816,[18] and Baron Saverbrun improved it in 1818. Because the price initially was high, the vehicle could be used only by the wealthy, including the British Prince Regent (1762–1830), subsequently George IV. The first true bicycle was built in 1839 by Kirkpatrick MacMillan of Scotland, a mechanic. He modified the Hobby Horse by adding a crank, pedals, driving rods, a seat, and handlebars. In 1846, Gavin Dalzell improved on MacMillan's machine. These two men were the pioneers of the bicycle.

Twenty years passed with little improvement until Pierre Lallement, an employee of M. Michaux of 29 Avenue Montaigne, Paris, brought forth the bicycle with a rotary crank. This vehicle could be commercially purchased at the listed address for 200 francs ($50 at that time). Immediately before the Franco-Prussian War of 1871, an extraordinary craze developed for these velocipedes. While the Coventry Sewing Machine Company was making small numbers of the *Michaux* for the English market, its French agent placed an order for 500 of this type. To handle this surge of orders, the company was reconstituted as the Coventry Machinist Co., Ltd., and later the Swift Cycle Co., Ltd.[19] Before the Michaux machines could be shipped, the war started. The company, faced with this stock of undeliverable items, created a market in England. These vehicles bore two almost equally sized wheels. The larger wheel was located in front, and the driving cranks and pedals were placed on it. Bicycles of this style were heavy, possessed wooden wheels with iron tires, and had a massive iron backbone. For their riding ability, they came to be known as "boneshakers."

With the demand for the machines rising, more makers offered improvements. The heavy wooden wheels gave way to lighter metal ones with wire spokes. With rubber becoming ever more common, solid rubber tires were glued to the rim. To enhance the stability of the two-wheeled velocipede, the three-wheeled tricycle was developed in 1850. At this time, the route followed by the bicycle and the tricycle diverged. The greater stability of the tricycle served as the frame for the development of both the electric-propelled and gasoline-powered vehicle. As shown in *History of the Electric Automobile: Battery-Only Powered Cars*,[2] Gustave Trouvé, in assembling the first electric vehicle, almost surely used the Dublin tricycle illustrated in Ref. 2.[20]

In contrast, Ayrton and Perry apparently employed either the Starley Royal Salvo tricycle available in 1880[21] or the Butler Omnicycle developed in 1879.[22] Each plate is shown in *History of the Electric Automobile: Battery-Only Powered Cars*.[2] Both vehicles had steerable small wheels. In converting, Trouvé and the two English professors apparently reversed the direction of normal operation so the steering wheel trailed rather than led. Riker appears to have modified the British Rudge 'Rotary' tricycle, which became available in 1880.[23]

Note that the early electric vehicle designers in the 1880s and 1890s converted what already existed. A similar path is being followed by the early builders of modern electric automobiles. As these automobiles are electrified, many are converting vehicles that exist, the gasoline-powered cars of the late 1970s, 1980s, and 1990s. An exception is the 1996 General Motors *EV1*. Large leaps in design occur only when an entirely new technique emerges, such as resulted in the airplane or in the nuclear reactor.

# Tires and Batteries

In parallel with electrical experiments, internationally the American, Charles Goodyear, was investigating the exudation from a tropical plant. It was to bear the name *rubber*, for in its perfected form it could be used to erase or rub out a pencil mark on paper.[24] The rubber source used by Goodyear was from the South American tree, *Manihot glaziovii*. Accidentally, but only after painstaking experiments, Goodyear learned that if this exudation were combined with sulfur and lead oxide in the presence of heat, the resulting product would be resilient and could be formed into useful articles of commerce such as a hose or tubing. (See Figure 1.1.) Goodyear invented the vulcanization of rubber in 1843.[25] Goodyear was the perfect example of the Biblical quotation: *One soweth, and another reapeth*, for he lived his life in poverty. Before and after his discovery, he was jailed for debt. Goodyear died in 1860.

*Figure 1.1. Charles Goodyear's discovery of rubber vulcanization in 1839 ushered in its commercial application. Although the discovery of making rubber commercially useful was a lucky accident, it was fortuity that only one who had experimented for years would recognize. After Charles Goodyear had tried hundreds of combinations for making rubber independent of the environment, in 1839 he dropped a piece of rubber mixed with sulfur on the kitchen stove. The combination of heat and sulfur gave the character that made rubber a desirable product. Goodyear patented vulcanization in 1844. (Ref. 25)*

In 1845, Robert W. Thomson had thoroughly and scientifically tested pneumatic tires on horse-drawn wagons; however, he lacked marketing knowledge and thus manufacturing ceased. Subsequently, a Scotsman named James Boyd Dunlop, a practitioner of veterinary medicine in Ireland, used inflated rubber tubing on his son's tricycle (shown in Ref. 2) and reinvented the pneumatic tire in 1888 in time for the explosive increase in bicycle use.[26] Seizing Dunlop's idea, the two brothers André and Edouard Michelin in France, whose firm manufactured solid bicycle tires in Paris, learned the technique of bolting pneumatic tires to the rim of a wheel. This technique was used with superior results by the brothers in the watershed 1895 Paris-Bordeaux-Paris automobile race.[27]

## The Lead-Acid Battery and the First Electric Cars

To propel an electric car requires a source of electrical energy. With sedulous effort, the distinguished French chemist Gaston Planté used two sheets of pure lead separated, it was said, by his wife's petticoat. In 1859 he immersed the assembly into a container bearing dilute sulfuric acid and developed the first lead-acid battery.[28] (Professor Planté is illustrated in Chapter 10 of Ref. 2.) This battery, with subsequent improvements by Faure, Sellon, and Julien, as more thoroughly described in Ref. 2, remains widely used in powering electric cars in the 1990s.

Also in France and just in time for the opening of the International Exhibition of Electricity, held in August 1881 in the Bois de Boulogne section of Paris, Gustave Trouvé, a trained experimentalist, exhibited the first electric personal vehicle, which was a tricycle.[29] He used lead-acid batteries and an electric motor of his own manufacture. (Trouvé appears in the Frontispiece of Ref. 2.) Observing this machine, Professors William Edward Ayrton and John Perry, on returning to their English laboratory, built an improved electric tricycle in 1882.[30] This tricycle is shown in Figure 1.2. Its speed was 9 mph (14.4 km/h), and its range was 20 to 30 miles (30 to 50 km) on hard roads, a performance somewhat better than a horse and buggy.

*Figure. 1.2. Ayrton and Perry's electric tricycle. (Scientific American. Ref. 30)*

# Generation and Distribution of Electricity

One year after the first electric car, the first central station for large-scale generation and distribution of electricity appeared. The year was 1882, and the facility was located on Pearl Street in New York City.[31] The power from this station would be delivered primarily to incandescent lamps developed in 1879 by Thomas Alva Edison. This generating station was peculiarly due to Edison, even to the design of the electric generators, switches, load control, etc. Its output was direct-current electricity, a mode of power that Edison long stubbornly held to be better than the alternating-current mode which Nikola Tesla and George Westinghouse favored and which today is nearly universally used when substantial demands for power are required.

At this time, Edison conceived another major concept, which was the professionally operated laboratory to develop commercial inventions. For his electricity generation, he also improved on Zénobe Theóphile Gramme's motor and generator. The latter enabled mechanical power to be converted into electric power on a massive scale when compared with energy available from a battery. From Edison's Pearl Street station, shown in Figure 1.3, issued a constant and large source of electric power. This station was the genesis of the web of electric generating stations throughout the world that contributes to today's standard of living. Figure 1.4 presents Edison in 1881.[32]

*Figure 1.3. Edison's Pearl Street station. (Scientific American. Ref. 31)*

*Figure 1.4. Thomas Alva Edison in 1881.* (Scientific American. *Ref. 32*)

Neither in France nor England did work on electric vehicles proceed evolutionarily. In contrast, entrepreneurs in America picked up the pace. Andrew L. Riker would build the first American vehicle, as outlined in Ref. 2. It was an unusual three-wheeled vehicle possessing a lead-acid battery and a d-c motor. The date was 1890.[33] (In 1899, 1.113 million bicycles worth $23 million were built in America. See Ref. 33.) The concept for personal electric vehicles had taken nine years to cross the Atlantic from France. Yet, in only four years, Henry G. Morris and Pedro G. Salom would form a company which in 1897 would first exploit electric vehicles commercially as taxis in New York City. Watching this, Colonel Albert A. Pope, a former Civil War cavalryman, with his Pope Manufacturing Company as the first major American bicycle maker (1877),[17] foresaw with his young MIT-graduated engineer named Hiram Percy Maxim that electric cars would be faster, cheaper, and cleaner than a horse and buggy.[a] With this prescience, Pope's company began assembling test electric cars.

---

[a] In 1900, New York City contained an estimated 300,000 horses. While you read this page, those animals dropped 10,000 lb (4,536 kg), or 5 tons (4.5 metric tons), of manure in the streets or stables.

# Early Transportation Problems

After the steam road engines had captured an ever larger share of passenger business from the horse-drawn omnibuses, why were the keen minds of England inactive in over-the-road vehicle development? For this impasse, the heavy hand of government had intervened, and the so-called "Red Flag Laws" were passed. In fact, there were four acts devoted to the "locomotive" (automobile): one each in 1861, 1865, and 1868, and sections of the Local Government Act of 1878. As defined in the acts, "locomotive" meant any vehicle propelled by any power except animal. In England prior to 1896, if a person wished to use a motor vehicle, or a "locomotive," at least three men were necessary and a fourth man was required to walk in front, bearing a red flag. In addition, the driver must obey a 2-mph (3-km/h) speed limit in town and a 4-mph (6-km/h) speed limit in the country. Furthermore, a license fee of £10 ($50 at that time) was required for driving through each county. If the above were insufficient, the vehicle could cross a bridge only if the latter bore a sign so permitting. Most of this restrictive regulation was swept away by the Fourth Locomotive Act of 1896.[34] However, the damage had been done. The legislation had hobbled the genesis of the English automotive industry.

These personal transportation laws, passed at the behest of the well-entrenched horse livery interests and the intercity horse-drawn stage coaches, were felt necessary, for discriminatory tolls alone were insufficient of a detriment. The steam road vehicle could race along in good weather and on improved roads at speeds as great as 40 mph (64 km/h) if permitted, a sustained speed much higher than any horse or coach could travel.[35] (On 13 July 1888, J. Selby drove the coach *Old Times* from London to Brighton on the south coast of England and return at an average speed of slightly under 14 mph (22.5 km/h), using eight teams and fourteen changes. See Ref. 35.) Indeed, there are examples of the British steam road vehicles passing a steam train on a level track. As early as 1831, discriminatory legislation existed. Table 1.1 shows tolls in 1831 between cited English cities, where the British tolls have been translated to American funds of the time.

**TABLE 1.1**
**TOLLS BETWEEN ENGLISH CITIES, 1831**

|  | Coach | Steam |
|---|---|---|
| Liverpool to Prescott | $1.00 | $12.00 |
| Bathgate Road | 1.25 | 6.75 |
| Ashburnnam to Totners | 0.75 | 10.00 |
| Feignmouth to Dawlich | 0.50 | 3.00 |

Such stifling, special-interest legislation in favor of horse-drawn Berlins and carriages stunted for 30 years the development of mechanical transport on English highways. Indeed, Parliament did not repeal the Red Flag Laws until 1896, 14 years after the first electric English vehicle. By that time, the Benz gasoline-powered vehicle had been well received in Germany, and the like-powered Panhard automobile was firmly established in France. Indeed, the 705-mile (1134-km) Paris-Bordeaux-Paris watershed automobile race had been run the previous year. The British politicians, abrogating the teachings of their own Adam Smith, the *laissez-faire* economist, severely damaged their country by over-regulation.

While personnel in the *Sceptered Isle* were inactive, the Germans, French, and Americans were producing a variety of motor cars: electric, steam, and gasoline-powered. Two examples of French steam road wagons of this early time are shown in Figures 1.5 and 1.6.[36] Figure 1.7 demonstrates an American gasoline-powered car of 1895.[37] Thus, at the turn of the century,

*Figure 1.5. Amédée Bollée's* père et fils *steam La Maucelle* of 1878 set the classic pattern of the automobile with front-mounted engine, transmission beneath the passenger compartment, rear wheels driven through a differential, and independent front suspension. (Ref. 36)

three alternate modes of personal transportation coincided competitively: the electric vehicle, the gasoline-powered automobile, and the steam-powered car. Races were held, and at that time, any of the three might win.

What was the opinion of Thomas Alva Edison in 1895 relative to this competition? At the time, Edison was widely considered the most important person in America, and fortunately a newspaper of the time recorded his remarks in an interview:[38]

> Talking of the horseless vehicle, by the way, suggests to my mind that the horse is doomed, yet this animal shows a greater economy of force than man, for 70% of the energy of the horse is available for work. But the horseless vehicle is the coming wonder. The bicycle, which ten years ago was a curiosity, is now a necessity. It is found everywhere.
>
> Ten years from now you will be able to buy a horseless vehicle for what you would have to pay today for a wagon and a pair of horses. The money spent in the keep of the horses will be saved, and the danger to life will be much reduced.

*Figure 1.6.* L'Obéssante, *1873, was a compact and well-proportioned vehicle. Père Bollée probably pioneered the flush-sided body on a true perimeter chassis frame, as well as independent front suspension. (Ref. 36)*

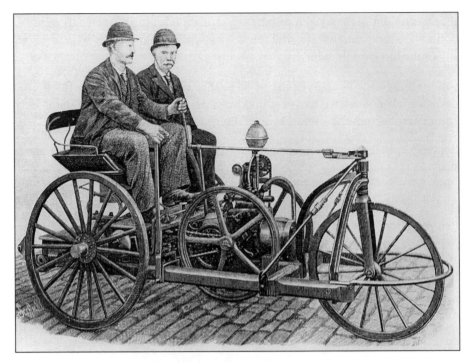

*Figure 1.7. An 1895 2-hp gasoline-powered tricycle motorcar built by A. Schilling & Co. in San Francisco. (Scientific American. Ref. 37)*

Will these vehicles be run by electricity?

I don't think so, said Mr. Edison. As it looks at present, it would seem more likely that they will be run by gasoline or naphtha motors of some kind. It is quite possible, however, that an electrical storage battery will be discovered which will prove more economical, but at present the gasoline or naphtha motor look more promising.

It is only a question of short time when the carriages and trucks of every large city will be run with motors. The expense of keeping and feeding horses in a great city like New York is very great, and all this will be done away with, just as the cable and trolley cars have dispensed with horses.

You must remember that every invention of this kind which is made adds to the general wealth by introducing a system of greater economy of force. A great invention which facilitates commerce enriches a country just as much as the discovery of vast hoards of gold.

# The Modern Approach

As you know, Edison's prediction proved correct. Then, as often happens, new minds, free to think, conceived of alternate means of controlling electricity. As a result of wartime research, William B. Shockley, John Bardeen, and Walter H. Brattain ushered in the solid-state electronic era with the discovery of the transistor in 1948. This event would have an impact. Subsequently, with the solid-state thyristor invented by Nick Holonyak,[b] the modern electric vehicle was awaiting birth. The commercial availability of solid-state current elements, combined with well-developed electric motors, improved batteries, chassis, and bodies, inaugurated a crop of electric vehicles. It was found that computer-designed electric cars could compete in speed and acceleration with gasoline-powered vehicles, yielding only in range. If confined to cities or resort use, their range would satisfy the perceived use of many drivers.[39] However, the limited range of the electric car and the discovery of *touring* in 1895 by Hiram Percy Maxim brought forth the hybrid electric car. A century later, the air pollution of our cities, largely from the gasoline-powered automobiles, was the impetus for the resurgence of interest in electric vehicles.

# Notes

Regarding Goodyear's invention of rubber, in the American Civil War the troops of the Union Army were issued rubber squares which enabled soldiers to be insulated from wet earth. In the event of pell-mell retreat of Union forces when soldiers abandoned items, as occurred at Virginia's Battle of Bull Run in 1861 (Manassas to the Confederacy), or at Tennessee's Battle of Chicamauga in 1863, these rubber squares were rich booty to the less well-equipped Confederates.

In 1888 in Ireland, shortly after the American War Between the States, the Scotsman James Boyd Dunlop[c] applied pneumatic rubber tires to his son's tricycle as shown in Figure 27.2 (page 430) of Ref. 2. Figure 1.8 shows Dunlop in later life. His experiment initiated the soon-widespread use of pneumatic tires on bicycles. Later, the Michelin brothers of Paris applied pneumatic tires to the motorcar that they entered in the watershed Paris-Bordeaux-Paris race of 1895, an event described in the Prologue of Ref. 2. This impressive 705-mile (1134-km) race of June 1895 was won, with repairs and all, at an average speed of 14.4 mph (23 km/h), and triggered the rapid development of this vehicle in the principal industrial nations at the time. The automobile was a wanted commodity!

The next chapter chronicles the development of petro-electric multi-powered automobiles.

---

[b] Holonyak did his research at the General Electric Company. Today (1996), he is a professor of electrical engineering at the University of Illinois, Urbana-Champaign.

[c] Dunlop reinvented the pneumatic tire. Robert W. Thomson had thoroughly and scientifically tested pneumatic tires on horse-drawn road wagons in 1845. The world simply was not ready.

*Figure 1.8. John Boyd Dunlop reinvented the pneumatic rubber tire in 1888, an item encouraging the bicycle and the subsequent automobile. (Scientific American)*

# References

1. Wakefield, Ernest H., *The Consumer's Electric Car*, Ann Arbor Science Publishers, Ann Arbor, MI, 1977.
2. Wakefield, Ernest H., *History of the Electric Automobile: Battery-Only Powered Cars*, Society of Automotive Engineers, Warrendale, PA, 1994.
3. *Experimenta Nova Magdeburgica*, Book 4, Jansson, Amsterdam, 1672, p. 147.
4. Letter to Dr. Lieberkuhn, 4 November 1745; Kruger, H.G., *Geschichte de Erde*, Luderwaldischen, Halle, 1746.
5. *Philos. Trans. R. Soc. London*, **47**, 565, 1751–1752.
6. *Bononiensi Scientiarum et Artium Instito Atque Academia Commentarii*, **7**, 363, 1791.

7. *Philos. Trans. R. Soc. London*, Part 2, **90**, 403, 1800.
8. Faraday, Michael, *Experimental Researches in Electricity*, reprinted, J.M. Dent & Sons, Ltd., London, 1912.
9. Tesla, Nikola, *Lectures, Patents, Articles*, Nikola Tesla Museum, Beograd, 1956.
10. Ampère, André Marie, *Essai sur la Philsophie des Sciences*, 2 V., Bechelier, Paris, 1838–1843.
11. Ohm, Georg Simon, *Das Grundgesetz des Electrischen Stromes*, Akademische Verlag Gesellschaft, 1838.
12. Pecqueur, Charles, *Des Ameliorations Materielles dans leur Rapports avec la Liberté*, 2nd ed., Gosselin, Paris, 1843.
13. Maxwell, James Clerk, *An Elementary Treatise of Electricity,* Garnett, William, ed., Clarendon Press, Oxford, 1881.
14. Hertz, Heinrich, *Electric Waves*, Jones, D.E., tr., MacMillan, London, 1893.
15. *The New York Times*, 19 January 1978, p. 31.
16. Caunter, C.F., *The History of the Development of Cycles*, Part I, Her Majesty's Stationery Office, London, 1955.
17. *Encyclopaedia Britannica*, Vol. 3, p. 544, Chicago, IL, 1959.
18. Caunter, C.F., *The History of the Development of Cycles*, Part I, p. 4, Her Majesty's Stationery Office, London, 1955.
19. *Encyclopaedia Britannica,* Vol. 3, p. 543, Chicago, IL, 1959.
20. Caunter, C.F., *The History of the Development of Cycles*, Part I, Plate IV, p. 11, Her Majesty's Stationery Office, London, 1955.
21. *Ibid.*, Plate V, p. 26.
22. *Ibid.*, Plate II, p. 7.
23. *Ibid.*, Plate V, p. 26. See also: Sharp, Archibald, *Bicycles and Tricycles*, 1977 MIT Reprint, 1896.
24. *Oxford English Dictionary*, Clarendon Press, Oxford, 1970, p. 855.
25. Pierce, Bradford K., *Trials of an Inventor: Life and Discoveries of Charles Goodyear*, Phillip & Hunt, New York, 1866.
26. du Cos, Arthur, *Wheels of Fortune*, Chapman Hall, London, 1938, p. 56.
27. *Scientific American Supplement,* **XXL**, 1023, 10 August 1895, p. 16343.
28. Planté, Gaston, *The Storage of Energy*, Elwell, P.B., tr., Whittaker & Co., London, 1887.
29. *Scientific American,* **XIV,** 362, 9 December 1882, p. 5767.
30. *Scientific American,* **XLVII,** 22, 25 November 1882, p. 343.
31. *Scientific American*, **XLVII**, 9, 26 August 1882, p. 127.
32. *Scientific American Supplement*, **XII**, 309, 3 December 1881, p. 4919.
33. Catlin, George B., "The Story of Detroit," *The Detroit News,* Detroit, MI, 1923.
34. *The Horseless Age*, **2**, 1 November 1896, p. 22. See also: *English Commons' Journal*, CLI, 1896, p. 58.
35. *Encyclopaedia Britannica,* Vol. 7, p. 665, 1956.

36. Barker, Ronald, and Harding, Anthony, tr., *Automobile Design,* Society of Automotive Engineers, Warrendale, PA, 1992, p. 25.

37. *Scientific American,* **LXXII,** 2, 12 January 1895, p. 25.

38. *The New York World,* 17 November 1895.

39. Schwartz, H.J., *The Computer Simulation of Automobile Use Pattern for Defining Battery Requirements for Electric Cars,* NASA TMX-71900, NASA Lewis Research Center, Cleveland, OH, 1976.

# CHAPTER 2

# The History of the Petro-Electric Vehicle

For more than four generations, experimenters have assembled electric vehicles bearing a supplemental energy source for the enhancement of range. Historically, this extra energy has been supplied by springs, flywheels, fuel cells, petroleum or natural gas-powered engines, and, most recently, solar cells. In a recent study, the U.S. Department of Energy identified 81 worldwide attempts in which one or two hybrid electric vehicles were made to yield extra range.[1] Therefore, we may conclude that range limitation has long been identified. To overcome this deficiency, two common approaches have been exploited: 1) to develop batteries with greater specific energy, and 2) to proceed with the hybrid principle.

The questions of batteries are discussed in Chapters 10 and 11 of *History of the Electric Automobile: Battery-Only Powered Cars*[2] and will receive little treatment here. Figure 2.1 illustrates two types of hybrid electric vehicles. The drawing to the left is a *series* type of hybrid electric vehicle. It contains an engine-generator, batteries, and an electric motor, the torque from which drives the wheels. Energy from the fuel is continually being transformed and stored as chemical energy and potential electric energy in the batteries. In contrast, with the *parallel* hybrid electric system, the petro-engine may uniquely drive the wheels, or together or separately the electric motor supplies torque to the wheels.

In general for both series and parallel hybrid electric vehicles, the supplemental energy source may be any of the five sources cited above. Years ago, I discussed the most recent, solar assist.[3] Because series and parallel systems both have possible economic niches, if indeed hybrid electric vehicles are ever viable, both series and parallel types may be found in service.[4]

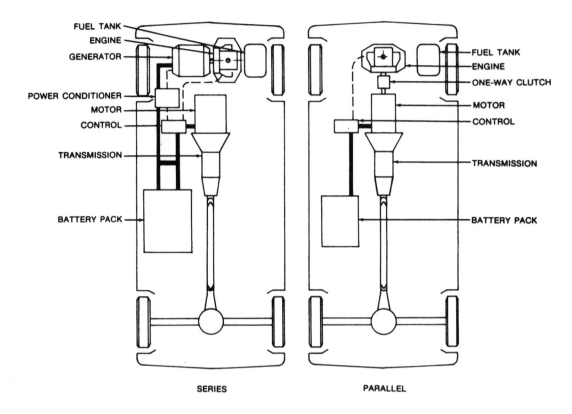

*Figure 2.1. In a series hybrid electric car (left), the engine drives a generator charging a battery. In a parallel system (right), the engine may directly couple the wheels. (Briggs & Stratton, 14 August 1995)*

## Competition for Nascent Power Systems

Only four years after Trouvé's original electric vehicle, described in Chapter 1, the initial gasoline-powered vehicle was tested in 1885. While 55 years would elapse from the operation of Thomas Davenport's electric motor shown in Ref. 2, the first to perform useful work,[5] until Trouvé's electric tricycle, the time span from N.A. Otto's four-cycle internal combustion engine of 1867 and a vehicle so powered was only 18 years.

From 1890 until 1905, a period of approximately 15 years, electric vehicles, internal combustion powered automobiles, and steam cars were highly competitive in America. On the other hand, in Europe, internal combustion cars were preferred from the beginning for many reasons: 1) the well-built Roman roads linking cities already existed in many parts of western Europe and thus range was important; 2) military-inspired highways were well maintained; 3) in contrast, roads linking American cities were largely unimproved and, with long periods of impassability, railroads were almost the only form of American inter-city travel; 4) moreover, the less severe European climate was more forgiving of water-cooled internal combustion engines; 5) Edison, Tesla, and Westinghouse gave later-evolving America a leadership in electric power, while Europe was more oriented to earlier developed steam power; 6) with a range sufficient for the small American cities of that time, the electric vehicle could flourish; and 7) possibly, too, with less regulation always present in a frontier-inspired society, entrepreneurship could more readily prosper.

Although an all electric vehicle could potentially displace only a small percentage of a petroleum-fueled personal vehicle market, it is presently perceived as occupying a restricted niche. On the other hand, the hybrid electric automobile is less subject to range limitation. With as little as 20% of the petroleum fuel of an internal combustion automobile, a hybrid electric vehicle may be discerned as viable.[6] The market is expected to make its own judgment. As the philosopher George Santayana wrote[7]:

"Those who know not the past are condemned to repeat it."

Therefore, what does history relate about the petro-electric drive system, two sample vehicles of which were ordered by the U.S. Department of Energy under the 1976 Electric Vehicle Act, and the many petro-electric cars for which a purchase order was issued by the City of Los Angeles in 1992?

The gasoline-electric car, currently referred to as the petro-electric vehicle, initially was an attempt to combine the high efficiency of the electric drive train with the remarkable energy storage of petroleum fuels. Presently, electric cars have a drive train efficiency of 90%, whereas a similar figure for the internal combustion automobile is probably 15%—a factor of nearly six in favor of electricity. As for fuel of both systems, the lead-acid battery in its storage of electric energy requires 1 kg (2.2 lb) for 125 Btu.[a] On the other hand, 1 kg (2.2 lb) of gasoline provides 45,300 Btu—a factor of 360 in favor of gasoline. When relative drive-train efficiencies are considered, there remains an enhancement factor of 70 to 80 in favor of gasoline. That large number almost shouts the reason why internal combustion cars will be here for many years. However, muting the above statement, and particularly in the last decade of the twentieth century, is the growing air pollution of urban areas largely as a result of the ubiquitous internal combustion vehicle. The petro-electric vehicle, the thought goes, could be petro-powered in country operation and electrically powered in urban regions, with the parallel type, hybrid electric automobile. If of series construction, the vehicle might have a low-polluting, constant-speed internal combustion engine driving a small generator that is continually replenishing the battery. Power from the battery would supply the electric motor.

# The First Petro-Electric Vehicle

Almost surely the first petro-electric car was built by Justus B. Entz, chief engineer of the Electric Storage Battery Company of Philadelphia. The date was 4 May 1897, as the chalkboard

---

[a] A common, 60-lb (27-kg) lead-acid automobile battery will emit approximately 1000 watt-hours. Thus, 1 kWh = 3413 Btu; 3413 Btu ÷ 27 kg = 125 Btu/kg.

drawing in the Pope Manufacturing Company indicates in Figure 2.2. His concept was not the simple approach of the early European designers of an internal combustion engine powering a generator which in turn charged a battery that drove the motor—a series type. As Hiram Percy Maxim, who early designed and built both electric and gasoline-powered automobiles, describes this vehicle:[8]

> Entz proposed a generator with a revolving field as well as a revolving armature. This (assembly) was made to act as a clutch. When the armature of this generator was short-circuited, its armature and revolving field became electrically locked together, and the engine drove the carriage through this locked clutch on what corresponds to high gear. When a grade was encountered, or for any other reason more driving torque was required, Entz eased off the short circuiting more or less, which permitted the clutch to slip. This slipping generated a current, and this current was fed to the electric motor, which was thereby enabled to help out the direct drive from the engine.

After this explanation, Maxim, as did many modern designers, wondered why all this electric gear was required when a simple 'gear box' (the transmission) could obviate the above and yield the drive system of a modern electric car.

In 1898, all was ready for testing the new car. Maxim, Entz, and a mechanic drove 'sluggishly,' the former relates, from the Pope factory to Capitol Street in Hartford, Connecticut. While on the street, Maxim left the vehicle for some reason and his foot caught on a wire to the ammeter. This breaking of the electrical circuit caused an electric arc to pierce the copper

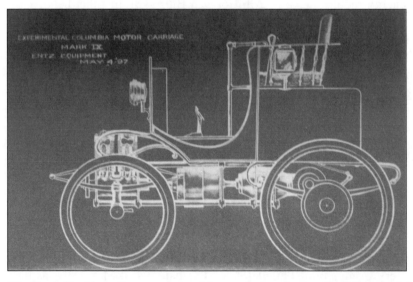

*Figure 2.2. Entz's blackboard masterpiece of the petro-electric car. (Ref. 8, copyright 1937 by Harper & Brothers; copyright renewed 1964 by Percy Maxim Lee; reprinted by permission of HarperCollins Publishers, Inc.)*

gasoline tank, and a stream of gasoline was ignited, burning the vehicle with the 7 gallons (26 liters) of contained gasoline. The Pope Manufacturing Company made no more petro-electric vehicles. However, Entz's idea did not die, according to Maxim. The concept eventually became the *Owen Magnetic Car.*

# The Belgian and the French Petro-Electric Cars

To implement the claimed advantages of the hybrid electric systems previously cited, both the Pieper establishment in Liège, Belgium, and the Vendovelli & Priestly Electric Carriage Company of France announced petro-electric vehicles at the Paris Salon of 1899.[9] In the Pieper vehicle, a small, air-cooled gasoline engine was forward and coupled to an electric motor powered by lead-acid batteries.[10] (See Figure 2.3.) Initially, the gasoline engine was started by the electric motor (probably among the first electric starters). When the vehicle was standing or coasting, the engine charged the battery. In contrast, when a hill was encountered requiring power greater than the air-cooled engine could provide, as the carriage slowed, additional electrical power flowed to the motor from the battery which provided supplementary torque to the rear wheels. In these early days, if the engine failed, and at this time all internal combustion engines were 'cranky' (a term used at the time), the reliable electric motor could be the power source to bring home the carriage!

*Figure 2.3. The Pieper* Stanhope. *(Ref. 9)*

21

The Pieper vehicle was of the parallel configuration. In contrast, the Vendovelli & Priestly carriage was of the series type and was three-wheeled. The two rear-wheels were each powered by a separate electric motor, the batteries being carried beneath the carriage to yield a range of 40 miles (64 km). For longer trips, a portable 308-lb (140-kg) engine-generator was added, consisting of a 3/4-hp De Dion-Bouton engine coupled to a generator capable of supplying continuously 10 amps at 110 volts (1.1 kW of power) to the battery. The engine/generator, the article stated, was also capable of supplying power sufficient to illuminate 15 10-candle (130-lumens) lamps, a phrasing used by the U.S. Army in purchase of its first electric vehicle in 1898 from the Woods Electric Vehicle Company of Chicago, as cited in Ref. 2.

Particularly unique to the Vendovelli hansom, however, was the steering mechanism. Steering was effected by utilizing the reversibility of the differential gears. By acting on the latter, one wheel revolved faster than the other, and the vehicle would turn with a short radius; indeed, the hansom can be made to pivot on itself by an action associated with the steering wheel before the driver. For braking the hansom, placing the controller in the *off* position also applies a friction-brake to a cast-iron pulley on the motor shaft. Figure 2.4 illustrates this French hansom. Figure 2.5 sketches the steering details.

*Figure 2.4. The Vendovelli & Priestley hansom.*
(Scientific American Supplement. *Ref. 11)*

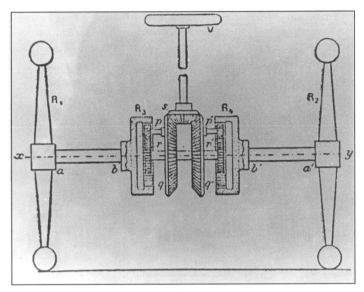

*Figure 2.5. Method of steering through the velocity of a variable wheel.*
*(*Scientific American Supplement. *Ref. 11)*

## Other European Hybrid Electric Vehicles

Because of the frequent impassability of American roads and the confinement of electric vehicles to the small cities of that time, all electrics had a better environment in which to thrive. This was not true in Europe where, for military reasons, urban areas were linked with well-developed highways, and inter-urban driving was routinely partaken by early drivers. While Camille Jenatzy, discussed in Chapter 15 of Ref. 2 on electric car racing, showed an early hybrid electric car at the Paris Automobile Show of 1901, little more was heard of it.[11] See Figure 2.6. In Jenatzy's car, the 6-hp engine could alone be used to directly drive the car, or it could be coupled to a 14-hp generator/motor to charge the batteries. These accumulators were stated to have capacity to deliver 18 kW of power. They weighed 370 lb (168 kg) and would have contained 4 kWh of energy. On electric power alone, the vehicle might be expected to have a range of 12 to 15 miles (19 to 24 km), a power consumption of 340 watt-hours per mile. With both engine and electric motor supplying power, it was said there was 20 hp at the wheels. The description of Jenatzy's automobile is seen as a parallel system.

There is seemingly a *Zeitgeist* in every industry, for the *Electric Times* of 1902 relates the French automobile trials for the promotion of alcohol consumption to help the Gallic wine growers.[b] For this competition, H. Krieger entered an alcohol-electric multi-powered car. His

---

[b] A solution of 95% ethyl alcohol and 5% water has a minimum boiling temperature and is thus not easily separated. At this ratio, the liquid is an *azeotrope*. *The Chemical Formularly* (1940) gives one patented French fuel with the following ratios: 83.4 ethyl alcohol, 13.9 water, and 2.8 denaturants. Being a champion of wine production, France has long thought in terms of indigenous alcohol-fueled vehicles.

*Figure 2.6. The 1901 Jenatzy petro-electric carriage.*
(Scientific American Supplement. *Ref. 11*)

vehicle was propelled by two Krieger slow-speed, four-pole, direct-current motors, each driving independently a forward wheel.[c] The motors were powered by 44 *Phoenix* lead-acid cells with a capacity of 120 amp-hr. Krieger's automobile also had a 4.5-hp alcohol-fueled engine coupled to a four-pole shunt-wound generator. When this vehicle traveled at 10 mph (16 km/h), the engine-generator balanced the current drawn from the battery supplying power to the drive motors. At greater speeds, the battery furnished the supplemental power. Curb weight of the car was 2600 lb (1179 kg). Krieger's automobile is recognized as a series type.

## The Lohner-Porsche Hybrid Electric Vehicle

In 1903, following the lead of Krieger, the Lohner-Porsche group had also developed a hybrid electric car based on the *Mercedes* gasoline car.[12] A 20-hp, four-cylinder engine drove a 21-kW, six-pole, d-c generator. A motor was in each of the two front wheels. The magnetic field of each motor was fixed, the 12-pole armature being a part of the wheels, free to turn. A protective cover for the motor was formed from the outer casting. The Lohner-Porsche was

---

[c] The General Motors *Impact* of 1990 revived this double-wheel drive, but the designers chose to use two 57-hp, variable speed, alternating-current induction motors.

a series system.  At that time, the builders used Planté's sheet-lead battery, which had the ability to allow heavy current draws for a short period, without flaking off the lead-oxide. The motor controller was under the center body and was worked by a lever beside the steering wheel.  The controller positions were:  battery alone to the motor, generator and battery, short circuiting the motor to act as a dynamic brake.  The radiator for the engine was in the front section.

# The *Auto-Mixte* Petro-Electric Car

With Jenatzy, Krieger, and now Lohner-Porsche building hybrid electric vehicles, in the 1906 Paris Automobile Show the *Auto-Mixte* series system car was shown.[13]  See Figures 2.7 and 2.8.  Notice the small gasoline tank.

> At the front (of the vehicle) is a four-cylinder, four-cycle gasoline engine. The gas throttle was operated by a magnetic device consisting of an iron-core working with a compound solenoid.  The latter opened the gas inlet to the motor to increase the speed should the voltage output of the generator fall, and *vice versa*.  Behind the generator, and seen in the insert, is a magnetic friction clutch consisting of the soft iron ring, C, which has two magnetizing coils sunk flush with the surface, and the iron disk, P, which is mounted opposite it and made to drive the shaft to the differential through a slidable universal joint. When the current from the battery flows through the coils, the attraction of the electro-magnet, C, for the disk, P, forms an easy and progressive clutch.

*Figure 2.7.  The* Auto-Mixte *series system car.  (Ref. 13)*

25

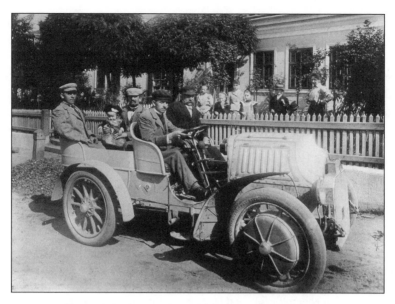

*Figure 2.8.  Dr. Ferdinand Porsche.  (Ref. 12, p. 173)*

The outer casing of the generator is extended farther in the rear, and on the other side of the disk, P, is mounted a second magnetic disk, B, which is fixed in the frame.  When the loose disk is attracted against B, a very effective magnetic braking action is obtained which can work in either direction.  The storage battery, A, of 24 cells and weighing 340 lb (154 kg) is placed under the driver's seat.  The system is completed by an electric controller which contains the circuit that allows the operation by a single lever:  to start the motor, to change the car speed, to stop the car, and to reverse the car.  The storage battery of the car is charged from the generator.  The car required no speed change gear, the generator, acting as a motor, starts the engine and, if the engine is shut off, the vehicle runs as an electric car.

Here the electric motor is portrayed to start the gasoline engine in the hybrid electric system.  The date is 1906 in France.  Earlier, Clyde J. Coleman had invented the electric starter in 1899.[14]  In 1912 in Dayton, Ohio, Charles Franklin Kettering commercially introduced the electric starter for internal combustion automobiles, an advance which substantially enhanced their acceptability.  The hand cranking of an engine, a source of many bone fractures, would become only a memory.  This feature was another nail in the coffin of the early electric car.  The apocryphal "little old lady" could now start and operate both electric and internal combustion powered cars.

## The *Mercedes-Mixte* Petro-Electric Car

Witnessing ever greater interest in hybrid electric vehicles, the Mercedes Company joined with Mixte in the 1907 Tenth Paris Automobile Show, and the resulting *Mercedes-Mixte* was shown as a petro-electric car. On the front of the chassis was mounted a four-cylinder, four-cycle gasoline engine of either 40 or 60 hp. In place of the flywheel behind the engine, a generator connected directly to the engine crankshaft. This generator rotor turned at the same speed as the engine and charged the battery. The output of the latter was delivered to the controller. No energy was consumed in resistance. The controller was located under the driver's seat and was operated by a lever at the side of the driver. The controller, by different voltage couplings, gave six speeds ahead and one in reverse.[d] Two electric motors were mounted on the rear wheels. Each motor housing formed the center structure of its wheel, the rim being the flange expansion of the motor. Use of the differential was thus avoided. See Notes at the end of this chapter.

## The Krieger Petro-Electric Car

On display in the same show, the Krieger gasoline-electric taximeter cab,[15] a later version of the one previously discussed, had mounted forward a four-cylinder, four-cycle gasoline engine direct-coupled to the motor shaft of the 24-hp, four-pole generator. See Figures 2.9 and 2.10. This dynamo supplied power to a storage battery. Power from the generator also supplied two 16- to 20-hp electric motors, each of which powered a front wheel through reduction gearing. "As regards the fare to be paid for the cabs, the mileage counter is worked by a flexible shaft from the rear wheel and registers \$0.15 for the first kilometer (0.6 mile/1.0 km), and \$0.02 for each 250 meters."

In Figure 2.9, the steel-enclosed controller can be seen behind the generator. The generator has a series and shunt winding (compound winding) and a third coil connected with the 40-volt storage battery which served to demagnetize the field, thus acting as a speed control.[16] The electric motors were on spring suspensions and ran at 1200 to 1600 rpm. This suspension served to ease the jerkiness on acceleration. Note the electric indicating meters, an early version of today's dashboard instruments. Indicating meters were first seen on the electric tricycle of Professors Ayrton and Perry in 1882, presented in Chapter 1. Both Krieger vehicles were series systems.

## Early American Petro-Electric Vehicles

After the gasoline fire destroyed the Entz-designed, Pope Manufacturing Company-built, petro-electric vehicle that was discussed earlier in this chapter, Knight Neftel constructed a hybrid electric automobile to compete in the New York–Boston Reliability Test starting 15 October 1902.

---

[d] Such a controller is described in detail in Chapter 12 of Ref. 2.

*Figure 2.9. Power plant of the Krieger car. The generator is seen at the left. (Ref. 15)*

*Figure 2.10. The Krieger petro-electric car. (Ref. 15)*

*Figure 2.11.  The Neftel petro-electric automobile.  (Ref. 17.)*

The Neftel machine, shown in Figure 2.11, was the only electric entry,[17] the others being either steam or gasoline powered.  Neftel's automobile, as well as most others, bore the presently employed torpedo design: engine/motor forward driving rear wheels, introduced by the Baker Electric Company of Cleveland.  Moreover, Neftel's vehicle was of the series petro-electric type.  Curb weight was 3500 lb (1588 kg).  A 10-kWh lead-acid battery, 64 cells with 75 amp-hr capacity, gave an electric power range of 15 miles (670 watt-hours per mile, which is high). When a hill was encountered, the extra power was furnished by the continually charged battery.  Other than being forced off the road by a horse-drawn wagon, Neftel's machine performed satisfactorily.  However, as is true of so many early experimenters in both America and Europe, Neftel appears for a priceless period then slips from sight forever.

## The Baker and the Woods Petro-Electric Vehicles

To resolve the range deficiency of battery-only powered cars, Baker and Woods, both well-known battery-only electric car manufacturers, in 1917 independently developed petro-electric automobiles.  Of this system, in *Scientific American* there appeared the "Woods' Dual."[18]

> An apparently successful effort to combine the advantages of the electric automobile with those usually associated only with the gasoline propelled type is represented by a radical new design known as the "dual-powered" car.  In this vehicle (which is illustrated in Figure 2.12), the general body lines and ease of control are those commonly attributed to the

luxurious electric automobiles, while the speed and increased radius of action of a gasoline car are also incorporated in the design. The power plant is a combination consisting of a 14-hp, four-cylinder engine mounted at the front of the frame in the usual position and a motor-generator placed immediately back of the engine. A magnetic clutch is interposed between them. A 24-cell (48-volt) storage battery is used, carried at the rear of the frame, but owing to the combination with the gasoline engine, only one-half of the usual number of cells are provided...

It will be evident that the machine can be started, operated through its entire range of speed, and stopped with the simple motion of a thumb and forefinger.

The arrangement of the various units is such that the car can be driven entirely by electricity, solely by the gasoline engine, or by both simultaneously. The maximum speed of 35 mph (56 km/h) is a reasonable speed for a vehicle of this character. A good feature is that when driving with the internal combustion engine's power, all of the energy that is produced that is needed to drive the machine at a certain speed is conserved because the electric machine automatically becomes a generator and charges the batteries. If for any reason the fuel supply should become depleted, the current available from the battery will propel the car 30 miles (48 km) over average roads on the full charge. If a steep gradient is encountered, the two forms of motive power that may be used as a boosting action can be obtained by utilizing the electric machine as a motor. There is no danger of stalling the engine because, should the engine cease to function, the electric machine would crank it over at several hundred revolutions per minute and of course immediately start it again.

The electric devices are very similar to those formerly used in the single unit starting and lighting system which were very popular for a time, except the combined motor-generator is of sufficient power to turn over the engine and propel the car at the same time.

The combination of the two sources of power have made it possible to get 48 miles (77 km) (of) travel from 1 gallon (4 liters) of gasoline. Considering the size and weight of the car and the maximum speed obtainable, this figure is apt to be given doubtful consideration by the technically informed. As the engine is of low power, it is possible to drive the car 30 miles (48 km) on a gallon of gasoline. When it is explained that by doing this, sufficient surplus current is generated to charge the batteries for an additional 15 miles (24 km), it will be apparent that the figure is not overstated.

A marked feature that will be appreciated is the elimination of the charging methods ordinarily followed in all electric cars, whether the current is taken from a power or lighting circuit. With this construction, the battery charging source is an integral part of the car, and very little trouble and expense are entailed by battery charging and maintenance.

The weight of the machine is slightly more than 3,000 lb (1361 kg), whereas that of the standard electric of the same type would be over 1,000 lb (454 kg) more. Owing to the small size of the 24-cell storage battery, considerable space is available in the rear compartment to carry an extra tire and supplies.

Reversing is easily accomplished by means of a heel pedal carried at the front of the driver's seat. The general arrangement of the gasoline power plant is the same as would be followed in a light gasoline car. The drive from the end of the motor-generator shaft to the rear axle is through two universal joints and a tubular drive shaft. The arrangement is such that when the car is descending a hill, by using the electric machine as a generator, the speed of the car is reduced and at the same time considerable electricity is being generated which is stored in the battery until needed.

While this car marks a radical departure from accepted automobile practice, the manufacturers have so much faith in its efficiency that they have discontinued making the line of straight electric drive cars formerly produced and have concentrated their entire energies of a large organization on the dual-power car. (In 1917, the car sold for $2,950—much more than an equivalent internal combustion car.)

Both Woods and Baker abandoned the dual-drive principle. Because each was a graduate electrical engineer, both reached the conclusion, after experimentation and vehicle operation, that the hybrid electric approach added complexity, increased maintenance, introduced gasoline and oil, multiplied the weight, and ballooned the cost. Each gave up. Woods, probably the most articulate and literary person of the early electric automobile industry, continued manufacturing electric cars in Chicago until 1921 when his company succumbed to the inevitable. Baker, on the other hand, as cited in Ref. 2, shifted to building electric forklift trucks and electric motors, the company's operation today. In those days, air pollution was not even 'a cloud the size of a man's hand in the sky.' Figure 2.12 shows Woods' dual-powered 14-hp *coupé* of 1917. The shape of today's vehicle is evident: derived from Baker's *Torpedo* of 1902.[e]

*Figure 2.12. Woods' dual-powered 14-hp* coupé. *(Ref. 18)*

---

[e] See Chapter 15 of Ref. 2.

# The Jeffery Petro-Electric Automobile

Once born, the petro-electric concept refused to die. In 1912, the Thomas B. Jeffery Company of Kenosha, Wisconsin, a noted bicycle maker, announced its 1913 Model *Cross-Country Gasoline-Electric Motorcar.* This vehicle bore a 38-hp, four-cylinder gasoline engine connected by its drive-shaft to a motor-generator rated at 8 hp (6 kW).

Although the motor-generator was primarily present for both starting the engine and recharging the battery, its design and output is relevant to modern hybrid electric vehicle builders. Battery recharging was with 24 volts at 30 amps. The Jeffery vehicle for meaningful electric car operation would require greater battery capacity, but the in-line design of the motor-generator and the use of its inertial mass for an engine-flywheel illustrates a possible direction for later petro-electric designers. The venerable Jeffrey Company, an early leading competitor to the Pope Manufacturing Company in building bicycles, would become, through various reorganizations, the American Motors Corporation which, in turn, was purchased by the Chrysler Corporation in 1987. From the sparse account presented here, contrasted with the greater activity exemplified on the Continent with its superior inter-urban road network, it is clear that Europe became the leader in early hybrid electric vehicles.

# Notes

The Mercedes Company licensed the *Auto-Mixte* concept to Lohner-Porsche. In the military maneuvers of 1902, Archduke Franz-Ferdinand, heir to the Austria-Hungary throne (assassinated in 1914 in Sarajevo, thus initiating World War I) was so impressed with the smoothness of his ride in the petro-electric automobile that he wrote a letter to his chauffeur, Ferdinand Porsche, then an army conscript. See Ref. 19. For more on the Archduke, please read *Twilight of the Hapsburgs: The Life and Times of Emperor Francis Joseph,* by Alan Palmer (Grove Press, New York, 1994). In the nineteenth century, Jakob Lohner was the head of an old established firm of coach builders to the Austria-Hungarian court. Lohner, following the European development of gasoline-powered motorcars, of which the chassis were based on carriages of the period, hired Porsche (1875–1952, and seen in the Frontispiece), a graduate of both the Vienna Technical College and subsequently the Brown-Boveri Company, a world-class electromechanical works. Porsche's task was to design a silent electric carriage. "The trouble with most horseless carriages," Porsche told Lohner, "is the complexity of their transmission. All this business of shafting, bevel gears, and chains could be quite simply avoided if we used electric cables to carry the power to the place where it is needed, which, obviously, is at the driving wheels." In the Paris Salon of 1900, Porsche's product was shown—a petro-electric automobile.[19] The motor-generator was under the hood, providing electrical power to front-wheel-mounted d-c motors.

While Porsche continued his interest, particularly in racing cars, in 1910 he designed his Land-Train for the Austria-Hungarian army based on the Mixte system. "The train comprised a 'locomotive' tractor and anything up to ten trailers," D.B. Tubbs writes,[19] "The locomotive was fitted with a 100-hp petrol engine, driving all its own wheels by means of electric motors in the hubs, while current could also be fed to hub-motors in each of the trailers..." This land train was a complete success in bringing supplies to the Austria-Hungarian army on the Italian front over muddy roads in 1917. As is well known, World War I began in August 1914. On both the Western and Eastern fronts, the conflict soon evolved into trench warfare. The battle lines evolved into a stalemate but with fearful casualties. (At the First Battle of the Somme, after an extensive bombardment of enemy lines, the British had more than 60,000 casualties on the first day, with more than 19,000 killed. In the three-day Battle of Gettysburg in the American Civil War, the North and South had 50,000 casualties but far fewer deaths.) How could troops "go over-the-top," break through barbed wire, ever present machine-gun fire, and capture the enemy in their trenches? Toward the middle of the longer-than-four-year conflict, English personnel secretly invented what is now known as the *tank;* at that time, a code word (water tank) was employed to bewilder the enemy. After the British Commonwealth armies learned how to exploit this new weapon, it was found to be a winner. In contrast, the Germans were slower to develop the tank, designing and building only 20. With the help of the tank (and other factors, particularly near-starvation on the German home front), the war ended 11 November 1918. Its price had been deadly.

After World War I, it became known that General Eric Ludendorff of the German High Command in the German War Program of 1917 "allotted almost nothing for tanks, (believing) truck production to be far more important."[20]

During the intervening 19 years between World War I and World War II, the German High Command further developed the tank. In the latter conflict, Porsche was called to design his Tiger tank. He lost in design competition to the Henschel Company. The Porsche design became a mobile gun.[21]

The next chapter indicates American effort in hybrid electric vehicles in modern times.

# References

1. Surber, F.T., *et al.*, "Hybrid Vehicle Potential Assessment," #5030-345, Jet Propulsion Laboratory, California Institute of Technology, Pasadena, CA, 1980.
2. Wakefield, Ernest H., *History of the Electric Automobile: Battery-Only Powered Cars*, Society of Automotive Engineers, Warrendale, PA, 1994.
3. Wakefield, Ernest H., *The Consumer's Electric Car*, Ann Arbor Science Publishers, Ann Arbor, MI, 1977, p. 114.
4. "Lucas Gives Evaluations of E/HV Systems," *Electric Vehicle News,* **10**, 2 May 1981, pp. 18–19.
5. Davenport, Thomas, "Specification of a Patent for the Application of Electro-Magnetism to the Propelling of Machinery," *Jour. of the Franklin Inst.,* **20**, 1837, pp. 340–343. U.S. Patent 132, 25 February 1837. Also, "Notice of the Electro-Magnetic Machine of Dr. Thomas Davenport of Brandon, Near Rutland, Vermont," *Am. Jour. of Sci.,* **32**, Appendix, 1837, pp. 1–8.

6.  Shafer, William F., Commonwealth Edison Company spokesman (retired), personal communication.
7.  Santayana, George, *Reason in Common Sense,* Berger Evans, *Dictionary of Quotations*, Delacorte Press, 1968, p. 511.
8.  Maxim, Hiram Percy, *Horseless Carriage Days*, 2nd ed., Harper & Brothers, New York, 1937, pp. 144–149.
9.  *Scientific American Supplement,* **XLVIII,** 1235, 2 September 1899, p. 19795.
10. Pieper, H., "Mixed Drive for Automobiles," Patent Number 913,846, 2 March 1903.
11. "Paris Automobile and Cycle Show," *Scientific American Supplement,* **LI**, 1319, 13 April 1901, p. 21142.
12. Barker, Ronald, and Harding, Anthony, eds., *Automobile Design,* Society of Automotive Engineers, Warrendale, PA, 1992, p. 169.
13. *Scientific American Supplement,* **LXI**, 1569, 27 January 1906, p. 25133.
14. Sadler, John Buist, personal communication, 22 July 1992.
15. *Scientific American Supplement,* **1668,** 21 December 1907.
16. Wakefield, Ernest H., *The Consumer's Electric Car*, Ann Arbor Science Publishers, Ann Arbor, MI, 1977.
17. *Scientific American,* **LXXX**, 17, 25 October 1902, p. 271.
18. *Scientific American,* **CXVI**, I, 6 January 1917, p. 23.
19. Tubbs, D.B., "Ferdinand Porsche," Barker, Ronald, and Harding, Anthony, eds., *Automobile Design*, 2nd ed., Ch. 6., Society of Automotive Engineers, Warrendale, PA, 1992.
20. Moyer, Laurence, *Victory Must Be Ours,* Hippocrene Books, New York, 1995.
21. Lightbody, Andy, and Poyer, Joe, *The Illustrated History of the Tank*, Publications International, Ltd., Lincolnwood, IL, 1989.

# CHAPTER 3

# Recent Petro- and Natural Gas-Electric Car Systems

In the same way as the early electric vehicle builders had turned to multi-powered design to enhance the range of electric vehicles, so did some of the engineers and innovators in the 1960s and later. In 1894, Carli had used the energy contained in stretched rubber bands. Others had employed the assist offered by a gasoline engine. However, the extent of efforts to increase the range of electric vehicles is revealed by literature searches which have found more than 300 references.[1]

In the modern era, additional sources were utilized for compensatory energy. Beside the petro- or natural gas-engine, there could be fuel cells, battery-battery systems, the flywheel with turbine as a prime mover, hydraulic storage, and the solar cell. Built or building in the workshops of the 1970s, 1980s, and 1990s would be at least one of each system. In type, they would be parallel and series hybrid electric systems. Although early experimenters had proven more range could be gained by dual-powered systems, would the market acquiesce in their continued production? Historically, it did not. Would phoenix-like restorations fare better? To enrich our background on hybrid electric systems, a volume worth reading is cited in Ref. 2. One of the early modern petro-electric systems was developed by Minicars in 1971.

## The Minicars Petro-Electric Drive Train

Minicars, an American company, chose a petro-electric multi-powered vehicle. As R. Rhoads Stephenson *et al.* write:[3]

> The Minicars parallel hybrid was an adaptation of a previously developed battery electric vehicle which was modified by the addition of a heat engine for demonstration purposes.[4] Under EPA (U.S. Environmental Protection Agency) sponsorship, the hybrid was subjected to a series of emission tests, with modifications introduced as the program progressed to improve emission performance. In its final configuration, the power train contained a Chevrolet *Corvair* air-cooled, horizontally opposed, six-cylinder engine and Lear-Sigler 20-hp shunt-wound electric motor with both shafts connected as a common unit. This combination provided power to the rear wheels of the specially designed vehicle through a torque converter and two-speed automatic transmission. (Some) 640 lb (290 kg) of batteries were distributed throughout the vehicle. Except for the use of an automatic transmission, the Minicars vehicle is similar in configuration and operation to the Petro-Electric Motors car, as seen in Chapter 3.
>
> In the final version, a spring-damper system was used to prevent rapid throttle valve operation, and the difference between the throttle command and throttle valve position was sensed and used to control the field current of the drive motor.
>
> Tests were performed using the California seven-mode cycle for various engine air-fuel ratios and simulated vehicle weights. On the basis of the results, it was concluded that hybrid operation alone, while it permits a reduction in emissions, is not sufficient to meet the original 1976 emission standards. Since fuel economy performance was not stressed in the design of the system, poorer economy resulted in the system which was optimized for emission performance.

## Wouk's Petro-Electric Automobile

Surely the leader of the modern petro-electric car movement has been Dr. Victor Wouk of New York City.[a] Probably more than any single person, he perceived a potential for the multi-powered approach to electric vehicle drive and was instrumental in having the United States government program on electric cars include hybrid electric test vehicles, beginning in 1975.

Wouk's attention to hybrid electric vehicles was based on enhancing the range of the battery-only electric car, the reduction of emissions from the internal combustion engine automobile, and the conservation of petroleum fuels. To achieve this trinity, Wouk and his associates, notably Dr. Charles L. Rosen, developed what they called a *compound parallel hybrid*. Figures 3.1 and 3.2 show Wouk, his vehicle, and the drive system.[5] A conversion of a

---

[a] See, for example, the October 1997 issue of *Scientific American* (pp. 70–74) for his readable summary of work on, and market conditions for, petro-electric hybrid cars.

mass-produced automobile, Wouk's vehicle was a 1972 Buick *Skylark* capable of seating five passengers comfortably. His drive source, linked to the shaft, was a 15-hp, separately excited, d-c motor/generator, and a *Wankel* rotary internal combustion engine of Mazda *RX-2* design connected through a manual transmission to the car's differential as shown in Figure 3.2.

Wouk's parallel petro-electric car had a curb weight of 4100 lb (1860 kg), contained eight 12-volt automotive batteries weighing 440 lb (200 kg), obtained a top speed of 80 mph (129 km/h), accelerated from 0 to 60 mph (97 km/h) in 16 seconds, and possessed a range of more than 200 miles (322 km) at 60 mph (97 km/h). While early petro-electric designers paid little heed to exhaust emissions, air pollution from internal combustion automobiles was certainly recognized in the 1970s. Emissions in grams per mile of Wouk's petro-electric vehicle of 1974 were as follows:

| Pollutants | Wouk's Average | 1976 Mandated Levels |
|---|---|---|
| Hydrocarbons | 0.33 | 0.41 |
| Carbon Monoxide | 2.46 | 3.41 |
| Oxides of Nitrogen | 0.80 | 1.00 |

*Figure 3.1. Wouk's petro-electric vehicle at The Environmental Protection Administration Laboratories, Ann Arbor, Michigan. (Victor Wouk)*

*Figure 3.2. Drive-train components in Wouk's compound hybrid. (Victor Wouk)*

While admitting his vehicle was not of optimum design, a fact visibly evident in Figure 3.1 as an overload in the rear and as the power installation shown in Figure 3.3, Wouk had achieved his original goal of enhanced range, reduction in petroleum consumption, and reduced pollutants. This assessment allowed the U.S. Department of Energy to more finely attune test specifications for the parallel petro-electric car procured from the General Electric Company. General Electric had won the bid for building a state-of-the-art petro-electric automobile discussed later in this chapter. Wouk continued his educational process by writing a series of papers.[6–8]

For his petro-electric car, Wouk had used a large internal combustion engine as had experimentalists in the early 1900s. My approach, next delineated, was to employ a small gasoline engine running at constant speed to drive an a-c generator with an output that, while driving, constantly charged the batteries supplying the electric motor. The design was the series petro-electric van.

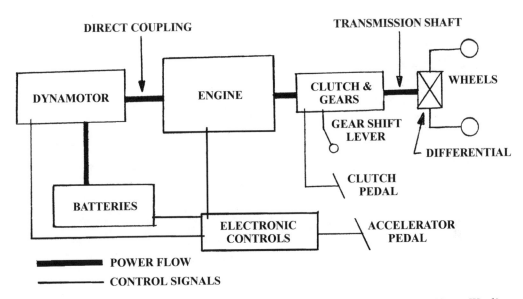

*Figure 3.3. Dynamotor and rotary engine installed in a petro-electric car. (Victor Wouk)*

# The Linear Alpha Petro-Electric Van

In 1967, I and my engineering group in Linear Alpha Inc. investigated a gasoline-electric drive system, with encouragement from the Illinois Bell Company and particularly John T. Trutter, a vice president. Elmer F. Yanke was the immediate liaison. For two approaches, the Onan and the Kohler engine-generator sets were procured. Each was rated 3 kW with an a-c output of 115 volts. Either electric energy system was to have a rectifier whose output fed batteries which in turn powered a pulse-width-frequency-modulated a-c drive system.[9] These experiments may have been the first self-contained petro-electric in which the drive motor was a-c powered. The complexity, increased weight, and added contamination of oil and grease of the petro-electric system arrested further progress. Because this petro-electric van would have a curb weight of approximately 5000 lb (2268 kg), a draw on the battery of 3 kW, which is the rate supplied by the generator, would have given the van a cruising speed of approximately 15 mph (24 km/h). A greater speed would require a net draw on the battery. The Linear Alpha a-c *Van 240,* as delivered, is shown in Figure 3.4. The vehicle was ultimately delivered to Illinois Bell as a battery-only a-c electric van being, it is believed, the first one of its type sold anywhere. Further information about this vehicle is found in Chapter 19 of *History of the Electric Automobile: Battery-Only Powered Cars.*[10] For this vehicle, helpful graphs, generated by Professor Gordon J. Murphy of Northwestern University, are found in Figure 3.5.[11]

A little later, a corporate giant would build a petro-electric vehicle.

*Figure 3.4. The final configuration of the Linear Alpha petro-electric vehicle. (Illinois Bell)*

## The General Motors Petro-Electric Vehicle

In 1969, General Motors in the United Kingdom developed the *512 Series* petro-electric car at almost the same time General Motors in Detroit was assembling the *Electrovair* and *Electrovan*, as reported in Chapter 19 of *History of the Electric Automobile: Battery-Only Powered Cars*.[10] The *512 Series* contained a 200-cm$^3$ engine coupled to a d-c series-wound electric motor through an electromagnetic clutch.[12] Taking advantage of the high starting torque of a series motor, in either mode, the vehicle accelerates from standstill as an electric car; in the petro-electric mode, the engine is engaged at 10 mph (16.1 km/h), at which speed the engine torque has reached a reasonable value. When the car is driven by the engine, a portion of the power can be used to drive the generator to recharge the battery. Top speed of the vehicle is 35 mph (56 km/h) in the petro-electric mode and 30 mph (48 km/h) in the electric. Range as a petro-electric vehicle is given as 150 miles (240 km) and, as an all electric, approximately 5 miles (8 km). Figure 3.6 is a line drawing of the *512 Series*, a sophisticated version of the Krieger petro-electric vehicle of 1907 shown previously in Figure 2.9.

In 1972, after a vehicle was destroyed by fire, the Electrical Engineering Department of the University of Illinois, Urbana-Champaign, ended its attempt with a hybrid electric car but succeeded in 1993 as demonstrated later in this chapter.

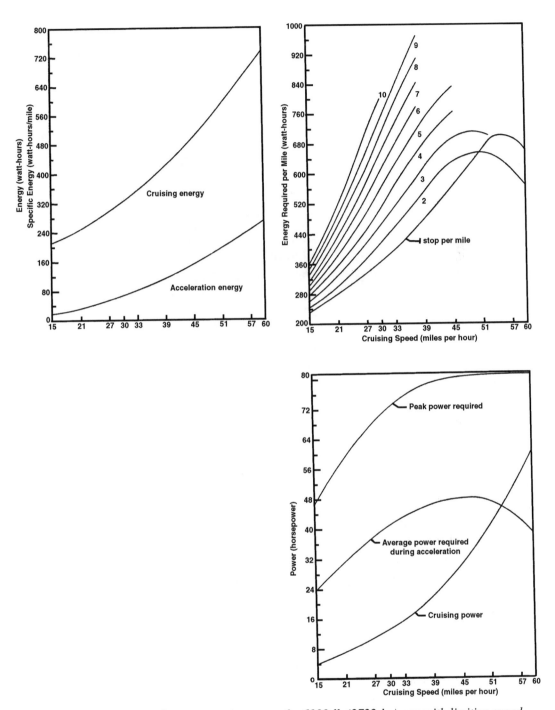

*Figure 3.5. Energy and power requirements of a 6000-lb (2722-kg) van with limiting speed of 65 mph (105 km/h) and initial acceleration of 5 mph (8 km/h) per second. (Ref. 11)*

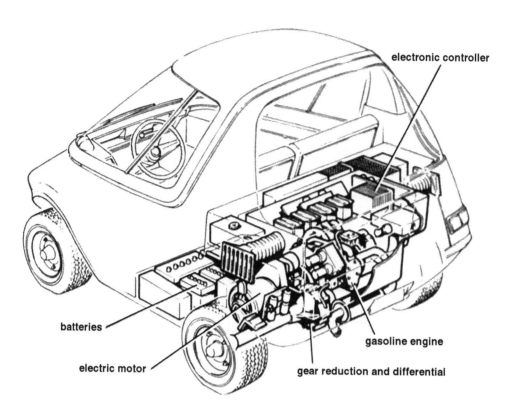

*Figure 3.6. The* 512 Series *petro-electric car is a neat example of space utilization.*
*(Ref. 12. Gillard Welch Ltd.)*

# Trailer Petro-Electric Automobiles

Another modern approach to the gasoline-electric car was first shown by Robert A. Aronson of Electric Fuel Propulsion Company of Detroit on 5 June 1975 in Washington, DC. The vehicle and trailer formed a single entity. The car was a converted Chevrolet *Chevelle*, renamed *Transformer II*, equipped with a 250-hp, d-c electric motor, powered by lead-acid-cobalt batteries of a nominal 180 volts. The motor was direct connected to the transmission, with power carried through the drive shaft to the differential and then the wheels. The 1850-lb (839-kg) trailer mounted the engine from the *Chevelle*, a 250-in.$^3$, six-cylinder, 105-hp unit driving a 37.5-kVA, 60-Hertz alternator operating at 1800 rpm. The generator was Y-connected, yielding 120 to 208 volts, capable of delivering 200 amps maximum. The voltage was rectified and controlled and delivered to the batteries. This series petro-electric system was driven from Detroit to Washington, DC, and return, making one overnight stop each way to allow the driver to sleep. Speeds as high as 65 mph (105 km/h) were reached, but the cruising speed was 50 mph (80 km/h). Probably the longest petro-electric vehicle run was made by the same manufacturer in delivering an electric car pulling a trailer from Detroit to an electric utility in Alaska. Figure 3.7 illustrates the trailer.

*Figure 3.7.  The 37.5 kVA mobile power plant can quickly recharge the  lead-cobalt batteries of* Transformer II *or, as a trailer, can constantly recharge while in motion.[13]  (Robert A. Aronson)*

At Linear Alpha Inc., of Evanston, Illinois, I also developed the trailer generator concept shown in Figure 3.8, to be used with the then d-c powered electric *Thunderbolt 240*, a converted Ford *Pinto* station wagon.  Although *Seneca* (without the trailer) reached Washington, DC, from its home factory, it arrived there by truck.  The concept proceeded only to the drawing stage, as illustrated.  The *Seneca* (without the trailer) is shown on the streets of Washington, DC, in Chapter 20 of *History of the Electric Automobile:  Battery-Only Powered Cars.*[10]

# The McKee Engineering Company Range Extender

The Achilles' heel of the electric motorcar has always been range.  To obviate this limitation, Robert S. McKee of the McKee Engineering Company of Lake Zurich, Illinois, a noted designer and fabricator of electric cars and Indianapolis 500-type race cars, made available to several electric motorcar manufacturers his version of the trailer approach.  As shown in Figures 3.9 and 3.10, the two-wheeled trailer carries an engine-generator to supply electric power to the battery in the electric vehicle.  This technique had been pioneered by Robert Aronson of Electric Fuel Propulsion Company in 1975, as cited in Figure 3.7.

McKee's approach was to employ a single-cylinder, four-cycle, 22.7-in.$^3$, 12.5-hp gasoline engine operating at 3800 rpm to drive a 7-kW d-c generator providing 180 volts.  With the trailer mechanically connected to the pulling vehicle and electrically linked to the power-supply battery, this entity, it is seen, leads to a series hybrid electric drive system.[14]

43

Now Possible To Go Cross-Country Without Stopping To Recharge

The Seneca Electric Vehicle can trav-el at 55 mph, and, in urban use, has a range of more than 30 miles with an experienced driver. Bearing an on-board charger the battery of the Seneca can be recharged from a 230 volt outlet. When connected to the Electric-Heart trailer, the Seneca can proceed cross-country. There is never need to stop and recharge the battery.

The Electric-Heart trailer is equip-ped with an engine-generator and gas-oline tank. With the Electric-Heart coupled to the Seneca as illustrated, pressing the start-button provides electric power to the battery of the Seneca enabling the driver to make extended trips. For local use the Electric-Heart is decoupled, and the Seneca can proceed on its own battery power.

The Seneca Electric Vehicle and Electric-Heart trailer enables the driver to draw energy for driving either from an electric outlet, or from a gasoline pump.

*Figure 3.8. The Linear Alpha* Seneca, *with trailer. (Linear Alpha Inc.)*

*Figure 3.9. Exterior of the McKee range extender. (Robert S. McKee)*

*Figure 3.10. Interior of the McKee range extender. (Robert S. McKee)*

# Aronson's Petro-Electric *Silver Volt*

In early 1982, Aronson's Electric Fuel Propulsion Company issued a modernized version of the trailer petro-electric car, designated the *Silver Volt*.[15]  Of this vehicle, Dr. Victor Wouk writes:[16]

> The manufacturer does not describe it (the *Silver Volt*) as a hybrid, although it is a parallel hybrid.  By the time you receive this letter, I should have driven in it from Los Angeles to San Diego nonstop.  It has already been driven in such a manner at 50 mph (80 km/h), with the batteries being depleted in a three-hour run over 150 miles (240 km) of driving.  For various reasons, the manufacturer is referring to it as an electric vehicle with range extension for emergency use.  When pressed, they will call it a "multiple fuel" vehicle rather than a hybrid.  They maintain that "hybrid" is too readily confused in the public mind, and the word "hybrid" is substituted.  This is not so far fetched, when I include herewith a copy of the page of the preliminary program of the IEEE session at which I am speaking...

**Electric And <u>Hybrid</u> Vehicles (A.2)**
Session Chairman: Dr. R.S. Kirk, U.S. Department of Energy

1. *Operational and Energetic Effectiveness of Road Electric and Hybrid Propulsion*
   G. Brusaglino, Centro Richerche Fiat

2. *PMC Electric Vehicle Controller*
   S. Post, PMC Electric Vehicle Components, Inc.

3. *Electric Vehicles: Practical With Today's Technology*
   V. Wouk, V. Wouk Associates

4. *Route Profile Analysis: Electric Postal Delivery Vehicles*
   C. Walter, M. Kong, D. Mullenhoff, and S. Wilson
   Lawrence Livermore National Laboratory

5. *Regenerative Characteristics: D-C Chopper Control Series Motor*
   M.H. Rashid, Concordia University

The *Silver Volt,* shown in Figure 3.11, was designed by Henry Lauve. This "top-of-the-line vehicle" is a five-passenger petro-electric with a reported top speed of 70 mph (113 km/h) and has a range of 80 to 100 miles (129 to 160 km) between charges. The fast-charge batteries are stated to be 80% charged in 45 minutes through the use of an onboard charger. If this vehicle consumed 400 watt-hours per mile, and if the range were indeed 80 miles (129 km), the batteries would require a capacity of 32 kWh and would have weighed 1900 to 2000 lb (862 to 907 kg). If the batteries were 80% charged in 45 minutes, the charging current at, say, 220 volts would have to be 200 amps. That is a high current to flow into a battery. With electric motor alone, the top speed is said to be 25 mph (40 km/h). Ref. 17 describes other characteristics. See also Figure 3.12.

# The Brobeck Petro-Electric Automobile

A most versatile entrepreneurial engineer has been William M. Brobeck of Berkeley, California. His activity in nuclear energy devices, in a steam-powered vehicle, had led him to electric multi-powered automobiles. This vehicle, unveiled at *Convergence '80*, features 15 12-volt marine batteries located in the trunk and a Westinghouse EA 40 shunt-wound, 0.09 lb-ft torque per ampere, 70 rpm per volt motor, which generates a maximum of 50 hp at 8000 rpm and 400 amps. Its electronic controls are located atop the batteries behind the rear seat. The gasoline engine is the standard 2.8-liter, four-cylinder, 90-hp motor supplied with the Chevrolet *Citation*. The 800 lb (363 kg) of batteries are said to provide the use of electric power for short trips without sacrificing the distance potential for long trips.[18]

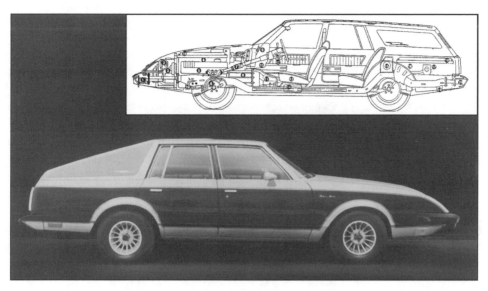

*Figure 3.11. The* Silver Volt. *(Electric Auto Corporation)*

*Figure 3.12. Outside his home in London, violinist Yehudi Menuhim (left) discusses his electric car with Robert Aronson on 8 September 1976. Aronson, from Detroit, was a pioneer in the renaissance of electric cars. Note the license plate designation. (Electric Auto Corporation)*

# Briggs & Stratton-Marathon Petro-Electric Automobile

A modification of Aronson's trailer petro-electric automobile was shown in 1980 by Briggs & Stratton Corporation of Milwaukee, Wisconsin, using the basic drive system of Marathon Electric Vehicle of Montreal, Canada.[19] Figure 3.13 pictures this six-wheeled car. The trailer, housing a 1000-lb (454-kg) package of lead-acid batteries, has been junctured with the body containing the torque-producing power plant. The vehicle is designed, it is stated, to carry two adults, two children, and groceries. As further illustrated in Figure 3.14, the shaft of the 18-hp engine interconnects with the shaft of the electric motor to drive the middle two wheels. This parallel petro-electric car accelerates from 0 to 30 mph (48 km/h) in 10.5 seconds in the electric mode to reach a top speed of 40 mph (64 km/h). With dual power, the speed reaches 55 mph (88 km/h). Using both power sources, the straightaway top speed is 75 mph (121 km/h). For additional specifications, see Notes later in this chapter. The Briggs & Stratton hybrid electric car was built as a technology demonstration, not as a production prototype. Note 3 at the end of this chapter provides additional specifications for this vehicle.

*Figure 3.13. The Briggs & Stratton parallel petro-electric automobile.*
*(Briggs & Stratton, 14 August 1995)*

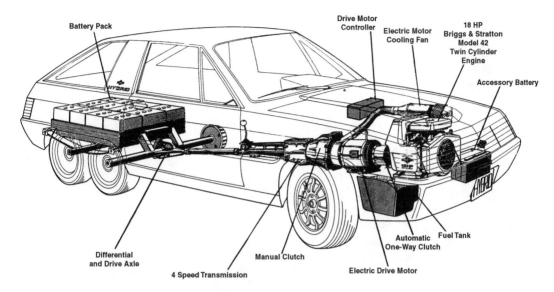

Battery Pack

Drive Motor Controller

Electric Motor Cooling Fan

18 HP Briggs & Stratton Model 42 Twin Cylinder Engine

Accessory Battery

Differential and Drive Axle

4 Speed Transmission

Manual Clutch

Automatic One-Way Clutch

Fuel Tank

Electric Drive Motor

*Figure 3.14. A cut-away of the Briggs & Stratton parallel petro-electric vehicle.*
*(Briggs & Stratton, see also Ref. 20)*

# The General Electric/U.S. Department of Energy Petro-Electric Automobile

The electric cars previously cited here were financed by the private sector as were the vehicles in the previous chapter. However, the Electric and Hybrid Vehicle Research, Development, and Demonstration Act of 1976 requires the building of hybrid electric vehicles financed by the public. The General Electric Company was chosen to develop a one-of-a-kind petro-electric automobile. This car, intended to seat five adults, has two drive systems: a 45-hp d-c electric motor and a 75-hp gasoline engine. The vehicle is calculated to consume 45 to 55% less gasoline than a conventional car when driven 11,000 miles (17,700 km) annually.[21,22] The prototype vehicle is shown in Figure 3.15; the drive system is illustrated in Figure 3.16.

In pricing the petro-electric vehicle, the U.S. Department of Energy contracted with five private firms for their versions of vehicle cost relative to its internal combustion counterpart. Emergent was a first cost of 21 to 62% higher. A positive payout over 10 years was, of course, a function of gasoline versus electric energy cost.[23] As with all costing of electric vehicles, there was controversy. Table 3.1 gives characteristics of several petro-electric vehicles of this period.[1]

*Figure 3.15. The General Electric Company HTV. (U.S. Department of Energy)*

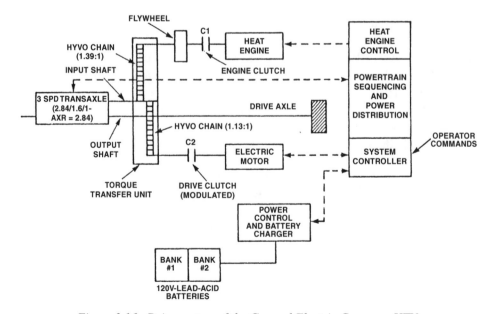

*Figure 3.16. Drive system of the General Electric Company HTV.*
*(U.S. Department of Energy, Contract #955190)*

## TABLE 3.1
## RECENT PETRO-ELECTRIC PASSENGER CARS

| | Briggs & Stratton | Brobeck Corp. | GE HTV[a] | Lucas Chloride | Volkswagen |
|---|---|---|---|---|---|
| Host Vehicle | Special 6-wheel | Chevrolet *Citation* | Special | Special | *Rabbit* |
| Curb Wt., lb (kg) | 3200 (1451) | 3920 (1777) | 4480 (2032) | 4630 (2100) | 2569 (1165) |
| Bat. Wt., lb (kg) | 792 (359) | 880 (399) | 970 (350)[b] | 1213 (550) | 441 (200) |
| Bat. Mass Fraction | 0.248 | 0.224 | 0.172 | 0.260 | 0.172 |
| Intermittent Motor Power, kW (hp) | 15 (20) | 45 (60) | 34 (46) | 50 (67) | 5 (7)[c] |
| Max. Engine Power, kW (hp) | 13.4 (18) | 67 (90) | 55 (74) | 30 (40) | 45 (60) |
| Range Bat. Only, km (mile) | 48 (30) | 40 (25) | <48 (30)[d] | 64 (40) | 36 (22) |
| Basis for Range | Not stated | At 72 km/h (45 mph) | FUDS[d] | UK urban conditions | At 52 km/h (32 mph) |
| Max. Speed, km/h (mph) | >89 (55) | >97 (60) | >90 (56) | 137 (85) | 145 (90) |
| References | (2) | (3) | (4) | (5) | (6) |

[a] General Electric HTV.
[b] 12.5 kWh lead-acid battery in car; JPL (Jet Propulsion Laboratory) dynamometer test results shown were obtained with a 72-cell Delco-Remy Ni-Zn battery weighing 726 lb (330 kg) for better economy and improved petroleum savings.
[c] Basis for rating not stated in reference.
[d] With peak loads over Federal Urban Driving Schedule (FUDS) supplied by ICE; battery charge was reduced to 20% SOC at 30 miles (48 km).
Source: Ref. 1.

# The McInnis Petro-Electric Car

Another innovative petro-electric vehicle in its drive system was developed by Stirling A. McInnis of Troy, Michigan, whose design, utilizing supplementary field coils, allows a variable weight of batteries to be employed. This feature gives the designer additional choices.[24]

# The 1993 Ford Hybrid Electric Vehicle (HEV) Challenge

One of the more effective means for developing a viable hybrid electric automobile has been initiated by the Ford Motor Company, with additional sponsorship by the Society of Automotive Engineers and the U.S. Department of Energy. Begun in the summer of 1993, the competition is patterned on the successful triennial (once sponsored GM Sunrayce USA 1990, now continued under the aegis of the U.S. Department of Energy) for universities and colleges. However, the Ford-sponsored competition is held annually and is limited to colleges and universities.

About the Challenge, the official bulletin relates:[25]

> The Ford Hybrid Electric Challenge is an intercollegiate competition that compels college students to look toward the future and develop designs for hybrid electric vehicles as a practical means of personal transportation. The student teams were given a choice of either converting a Ford *Escort* station wagon or constructing a hybrid from the ground up. A hybrid electric vehicle (HEV) has two sources of power used for propulsion—one or more electric motors and associated batteries, and an auxiliary power unit (APU). The APU may connect to the wheels directly and/or power a generator to make electricity for the electric motor(s). In the HEV Challenge, the APU is restricted to the use of denatured ethanol (E100), M85 (a blend of 85% methanol and 15% hydrocarbons), and reformulated regular unleaded gasoline.

> The participating teams will compete for awards in excess of $75,000. Awards will be given to the overall winners and to top finishers in each competitive event for both the conversion and ground-up classes. Additional awards will be given in a variety of categories, including design innovations, safety, environmental friendliness, team spirit, and best use of the HEV Challenge to generate interest in math and science among youths.

Table 3.2 lists HEV competitive events and scoring.

**TABLE 3.2**
**HEV CHALLENGE COMPETITIVE EVENTS AND SCORING**

| Dynamic Events | Date | Points | Static Events | Date | Points |
|---|---|---|---|---|---|
| Emissions | 6/3 | 150 | Engineering Design | 6/2 | 150 |
| Range | 6/4 | 75 | HEV Technical Report | 6/2 | 75 |
| HEV Commuter Challenge | 6/5 | 150 | Cost Assessment | 6/3 | 125 |
| Vehicle Efficiency | 6/5 | 125 | Oral Presentation | 6/3 | 50 |
| Acceleration | 6/5 | 100 | | | |

All vehicles must pass safety inspections and vehicle qualifying prior to competing in dynamic events (6/1 to 6/2).

Inasmuch as the details of this competition for petro-electric automobiles are related in *Innovations in Design: 1993 Ford Hybrid Electric Vehicle Challenge*,[26] presented here is a truncated version. To summarize the 30 competing vehicles in the Ford HEV Challenge:

A. Twelve vehicles were designed from the ground up, in the author's two volumes designated "compleat electric vehicles"; 18 were conversions.

B. Fifteen drive systems had parallel architecture, and 15 employed series architecture. (See Chapter 2 for definitions).

C. Twelve vehicles employed alternating-current drive; 18 were direct-current drive.

D. For batteries, 23 used lead-acid, 5 were nickel-cadmium, and 1 was nickel-metal-hydride.

E. System operating voltage varied from 72 to 394 volts.

F. Eighteen of the vehicles had regenerative braking.

G. Gross vehicle weight varied from 2346 to 4382 lb (1064 to 1988 kg).

H. With electric drive only, range varied from 15.6 to 54.8 miles (25.1 to 88.2 km).

J. Six university vehicles were either nonstarters or were disqualified.

Table 3.3 lists participants in the 1993 Ford HEV Challenge.

**TABLE 3.3**
**PARTICIPANTS IN 1993 HEV CHALLENGE**

### Conversion Vehicles

| Team | Vehicle No. |
|---|---|
| University of Alberta | 8 |
| California State University-Northridge | 86 |
| University of California-Irvine | 17 |
| Colorado School of Mines | 10 |
| Colorado State University | 4 |
| Concordia University | 66 |
| University of Illinois | 88 |
| Jordan College Energy Institute | 77 |
| Pennsylvania State University | 11 |
| Seattle University | 0.1 |
| Stanford University | 100 |
| Texas Tech University | 0 |
| U.S. Naval Academy | 96 |
| Washington University-St. Louis | 23 |
| Wayne State University | 99 |
| Weber State University | 6 |
| West Virginia University | 35 |
| University of Wisconsin | 110 |

### Ground-Up Vehicles

| Team | Vehicle No. |
|---|---|
| California State Polytechnic University-Pomona | 233 |
| California State Polytechnic University-San Luis Obispo | 101 |
| University of California-Davis | 30 |
| University of California-Santa Barbara | 56 |
| Cornell University | 42 |
| University of Idaho and Washington State University | 31 |
| Lawrence Technological University | 1 |
| Michigan State University | 55 |
| New York Institute of Technology | 22 |
| University of Tennessee | 33 |
| University of Texas-Arlington | 93 |
| University of Tulsa | 3 |

By the scoring system in Table 3.2, the right column of Table 3.4 gives the total point score for each participating school for both ground-up designed and conversion vehicles. In the former category, Cornell University is seen as the winner; for the conversions, the University of Alberta carried the honors. Noted should be the novel means by which the University of Alberta financed its vehicle: a cooperating local electric utility included in its billing envelopes a request that the recipient customer donate to the University petro-electric motorcar project.

# The University of Illinois Entry

To provide further credence, the experience of the University of Illinois, Urbana-Champaign, entry is used as an example. The 29 entries from other competing schools could also have been employed as a paradigm. As Dr. Philip T. Krein, Professor of Electrical and Computer Engineering, kindly wrote to me on 3 September 1993:

> As you know, the University of Illinois entered the Ford *Escort* conversion hybrid vehicle in the 1993 Ford Hybrid Electric Vehicle Challenge. We also expect to enter the car in the 1994 Challenge. The car was awarded first place in design for the conversion class and also received honors for Best Use of Materials and for Best Use of Electronics. It finished fifth overall, primarily because the generator control did not function as designed during the competition. (The completed car with all the sponsoring companies' logos is shown in Figure 3.17).
>
> The car is of the series architecture—it is basically an electric vehicle, with an auxiliary engine-generator set. The engine burns ethanol, gasoline, or any mixture (a manual adjustment sets it up for the intended fuel). Basic information, specifications, and measured performance are listed on the enclosure (Table 3.5). The vehicle was the only one at the competition to retain the complete stock interior, including all rear-end cargo space. The philosophy of our student team is to build a practical hybrid car, and the interior, the power-train layout, and the rather unassuming dashboard demonstrate their success in this direction. There are no displays or dashboard modifications other than the ones visible in the photos (Figures 3.18 and 3.19). We have not yet completed all the tests, and the hybrid range is still uncertain. Of course, hybrid range is limited mainly by fuel capacity, and we are refitting the car for a 10-gallon tank. This should allow it to duplicate the range of the original *Escort*.
>
> The batteries are fully enclosed in an aluminum box and are mounted outside the passenger compartment underneath the rear. In the enclosed photographs, the pack is completely invisible. The floor level and ground clearance were both maintained since the battery box simply takes the place of the fuel tank and spare tire well. A spare is still provided, mounted in the conventional station wagon location inside a rear fender.

**TABLE 3.4**
**SORTED OVERALL POINT TOTAL**
**U.S. DEPARTMENT OF ENERGY**

| Car Number | School | Vehicle Type | 15.9 Technical Report | 15.6 Engineering Design Event | 15.7 Oral Presentation | Penalty Points | 15.5 Acceleration Event | 15.1 Emission Event | 15.2 Commuter Challenge Event | 15.4.1 APU Efficiency Event | 15.3 Range Event | 15.4.2 Electric Efficiency Event | 15.4.3 Overall Efficiency Event | 15.8 Cost Assessment Event | Total Points |
|---|---|---|---|---|---|---|---|---|---|---|---|---|---|---|---|
| 42 | Cornell University | Grd-up | 37 | 44 | 41 | | 82 | 106 | 150 | 29 | 62 | 29 | 45 | 125 | 750 |
| 30 | University of California - Davis | Grd-up | 51 | 87 | 11 | | 79 | 77 | 102 | 24 | 75 | 24 | 37 | 108 | 675 |
| 55 | Michigan State University | Grd-up | 62 | 124 | 34 | | 41 | 99 | 124 | 20 | 51 | 17 | 19 | 60 | 651 |
| 33 | University of Tennessee | Grd-up | 43 | 22 | 29 | | 63 | 111 | 87 | 17 | 37 | 20 | 22 | 74 | 525 |
| 1 | Lawrence Technological University | Grd-up | 26 | 150 | 24 | | 41 | 68 | 52 | 8 | 26 | 8 | 13 | 66 | 482 |
| 53 | University of California - Santa Barbara | Grd-up | 75 | 36 | 14 | | 31 | 85 | 61 | 12 | 43 | 12 | 16 | 93 | 478 |
| 31 | University of Idaho/Washington State | Grd-up | 8 | 102 | 7 | | 84 | 48 | 44 | 10 | 31 | 14 | 55 | 44 | 447 |
| 3 | University of Tulsa | Grd-up | 22 | 29 | 50 | | 20 | 84 | 36 | 35 | 22 | 0 | 10 | 49 | 357 |
| 233 | California Polytechnic - Pomona | Grd-up | 18 | 52 | 9 | | 34 | 0 | 73 | 6 | 18 | 10 | 32 | 83 | 335 |
| 101 | California Polytechnic - San Luis Obispo | Grd-up | 14 | 73 | 20 | | 23 | 42 | 0 | 14 | 14 | 35 | 27 | 55 | 317 |
| 93 | University of Texas - Arlington | Grd-up | 11 | 61 | 17 | | 0 | 0 | 0 | 0 | 0 | | 0 | 25 | 114 |
| 22 | New York Institute of Technology | Grd-up | 11 | 0 | 5 | | 0 | 0 | 0 | 0 | 0 | | 0 | 25 | 41 |
| 8 | University of Alberta | Conv | 46 | 131 | 43 | | 88 | 71 | 150 | 24 | 75 | 14 | 38 | 94 | 774 |
| 6 | Weber State University | Conv | 16 | 93 | 50 | | 59 | 131 | 115 | 18 | 65 | 31 | 31 | 125 | 734 |
| 66 | Concordia University | Conv | 75 | 103 | 10 | | 78 | 82 | 76 | 35 | 34 | 9 | 48 | 101 | 651 |
| 0.1 | Seattle University | Conv | 28.5 | 61 | 25 | | 48 | 97 | 131 | 20 | 57 | 35 | 25 | 75 | 602.5 |
| 88 | University of Illinois | Conv | 65 | 150 | 31 | | 52 | 73 | 40 | 9 | 27 | 12 | 42 | 57 | 558 |
| 35 | West Virginia University | Conv | 42 | 76 | 22 | | 44 | 48 | 84 | 10 | 21 | 18 | 55 | 70 | 490 |
| 10 | Colorado School of Mines | Conv | 12 | 55 | 16 | | 51 | 71 | 68 | 31 | 46 | 11 | 18 | 81 | 460 |
| 96 | United States Naval Academy | Conv | 38 | 40 | 12 | -10 | 32 | 88 | 93 | 27 | 30 | 7 | 16 | 87 | 460 |
| 4 | Colorado State University | Conv | 34 | 84 | 34 | | 8 | 74 | 45 | 22 | 38 | 8 | 20 | 50 | 417 |
| 86 | California State University - Northridge | Conv | 18 | 50 | 28 | | 32 | 89 | 35 | 12 | 16 | 24 | 22 | 65 | 391 |
| 100 | Stanford University | Conv | 51 | 115 | 5 | | 24 | 0 | 61 | 0 | 18 | 27 | 12 | 47 | 360 |
| 110 | University of Wisconsin | Conv | 57 | 68 | 38 | | 68 | 0 | 0 | 0 | 14 | 18 | 0 | 41 | 304 |
| 77 | Jordan College Energy Institute | Conv | 14 | 35 | 18 | | 44 | 0 | 55 | 8 | 24 | 22 | 28 | 53 | 301 |
| 99 | Wayne State University | Conv | 28.5 | 30 | 14 | -240 | 46 | 106 | 50 | 16 | 51 | 10 | 34 | 87 | 232.7 |
| 11 | Pennsylvania State University | Conv | 0 | 25 | 8 | -327 | 56 | 67 | 103 | 14 | 42 | 18 | 14 | 61 | 81.2 |
| 17 | University of California - Irvine | Conv | 21 | 0 | 7 | | 0 | 0 | 0 | 0 | 0 | | 0 | 25 | 53 |
| 23 | Washington University - St. Louis | Conv | 0 | 0 | 6 | | 0 | 0 | 0 | 0 | 0 | | 0 | | 6 |
| 0 | Texas Tech University | Conv | 24 | 45 | 20 | -222 | 0 | 56 | 0 | 11 | 0 | | 0 | 44 | -21.5 |

*Figure 3.17. The University of Illinois petro-electric Ford* Escort. *A 22-hp, four-cycle engine drives an 8-kW generator supplying power to a frequency and pulse-width modulated system which in turn provides variable frequency power to a three-phase induction motor. Regenerative braking is provided. (Philip T. Krein, University of Illinois, Urbana-Champaign.)*

## TABLE 3.5
## SPECIFICATIONS AND CHARACTERISTICS OF
## THE UNIVERSITY OF ILLINOIS COLLEGE OF ENGINEERING
## (URBANA-CHAMPAIGN) HYBRID ELECTRIC VEHICLE

**Basic Outline:** The University of Illinois (UIUC) Hybrid Electric Vehicle was prepared for the Ford/U.S. Department of Energy/SAE 1993 Hybrid Electric Vehicle Challenge. It is a converted 1992 Ford *Escort* station wagon. The UIUC car was the only one in the Challenge to retain the complete five-passenger interior structure of the stock car.

**System Arrangement:** Series hybrid, with full a-c electric traction drive system. The electric system is supplemented by a small fueled engine-generator set.

**Summary of Major Components:**

| System Element | Description | Capability or Characteristics |
|---|---|---|
| Main drive motor | Three-phase a-c squirrel-cage induction motor | 90 hp peak at 5400 rpm; 0–8000 rpm speed range |
| Mechanical drive system | Stock Ford *Escort* arrangement, with elimination of clutch | Five-speed clutchless manual transmission, stock transaxle |

## TABLE 3.5 *(continued)*

| System Element | Description | Capability or Characteristics |
| --- | --- | --- |
| Engine for E-G set | V-twin fuel-injected four-stroke water-cooled engine. Fuel adjustment for ethanol, gasoline, or any mixture. | 22 hp at 3000 rpm |
| Generator for E-G set | Three-phase a-c squirrel-cage induction motor | 8 kW output continuous, up to 15 kW short-term, 9000 rpm |
| Inverter for electric drive | Constant volts per hertz system with slip control | 100 A continuous output; IGBT inverter system |
| Brakes | Dual regenerative and power-assist standard *Escort* brakes | Regeneration for all normal braking action. Mechanical brakes are applied when sufficient pedal travel appears |
| Batteries | 26 sealed glass-mat lead-acid units from computer backup supply application, connected in series | 312 V nominal (split for ±156 V), 8.8 kW·hr at 3-hour discharge rate, 307 kg |
| Battery pack | Enclosed flat pack mounted externally in place of stock fuel tank and spare tire | Fully enclosed plug-in pack system, with slide-in mount method. Dimensions: approximately 15 cm high, 80 cm wide, 145 cm long |
| Displays | Stock dashboard. Stock display (from Mazda *Protege*) with LED graph used to show battery state of charge | — |
| 12 V system | dc-dc converter from main battery bus | Up to 1500 W output, 12.0 V, fixed output as battery voltage decreases |
| Electronics | Microcomputer network with multi-task operating system, for displays, diagnostics, and battery monitoring | Complete vehicle control retained if computer network is nonfunctional |
| Charging interface | d-c port in place of fuel filler, with electronic safety interlock | Up to 30 A charge current at full d-c input |
| Charger | Switching power supply unit, garage mounted | Input source: standard 220 V or 208 V, 30 A or 50 A outlet. Output: d-c for battery pack, with interlock |
| Interior | No modifications to interior space except roll cage required by Challenge rules | Complete five-passenger interior space retained. Cargo space, seats, dashboard, and accessories per stock *Escort*. |

**TABLE 3.5 *(continued)***

**Performance Characteristics:**

| Parameter | Value | Test Basis |
|---|---|---|
| Acceleration | 100 m in 9.1 s (0.25 g); 0–50 mph in 9.3 s. | Measured at HEV Challenge competition |
| Top speed | 95 mph (estimated) | 92 mph achieved in early tests |
| Range, electric | 33 miles, 45 mph | Measured on highways near campus |
| Range, hybrid | >225 miles, 45 mph | Limited only by fuel tank size; batteries do not discharge in hybrid mode at 45 mph |
| Charging | 90% of charge restored in 4 hours from 30 A, 208 V outlet.; full charge in approximately 6 hours | Multiple charge cycles with full discharge |

**Other Characteristics:**

Curb weight: 3240 lb
Date of first operation: June 2, 1993

*Figure 3.18. Dashboard of the University of Illinois petro-electric* Escort.
*(Philip T. Krein, University of Illinois, Urbana-Champaign)*

*Figure 3.19. Front region of the University of Illinois petro-electric* Escort.
*(Philip T. Krein, University of Illinois, Urbana-Champaign)*

The 1993 Ford Hybrid Electric Vehicle Challenge is seen to reflect an increasing interest in the hybrid electric vehicle as the one most capable of giving sufficient range as expressed by the large body of users, and one in which infrastructure (filling stations) already present can be readily adapted for the required fuel.

## The Natural Gas-Electric Multi-Powered Automobile

An alternate to the petro-electric multi-powered motorcar previously cited is a modified internal combustion-powered vehicle fueled by natural gas. Unique Mobility Inc. U.S.A. has developed a drive system for a minivan to be used in the fleet of the San Diego Gas & Electric Company, U.S.A.

The series drive hybrid electric system consists of a Ford 1.3-liter, four-cycle, four-cylinder Otto engine configured to operate on natural gas,[b] capable of yielding continuously 25.4 kW at 2800 rpm to a 50-kW alternator, and this power to the battery pack. Without using battery power, this engine is capable of driving the van at 60 mph (97 km/h) on a level road. The

---

[b] Your car contains an Otto engine. Nikolaus August Otto (1832–91), co-inventor of the internal combustion engine, in 1876, improved the four-stroke Otto cycle.

battery pack consists of 15 Johnson Controls Company 12-volt, 55-amp-hr, sealed lead-acid batteries. Total weight of the battery is 600 lb (272 kg), thus eliminating approximately half the usual battery weight while giving the vehicle a range limited only by the amount of natural gas carried. The battery has the capability of delivering for a short time 80 kW of power to two 50-kW motors, each weighing 42 lb (19 kg). These motors, fed from an inverter-controller, are essentially six-phase wound, permanent magnet synchronous motors of original design. Each motor drives one of the front two wheels through a 5 to 1 reduction planetary gear to a sprocket and chain axle drive.[27] The drive assembly is shown in Figure 3.20.

Yet another natural gas-Otto engine-generator-battery vehicle is one sponsored by the Grumman Corporation and the Long Island Lighting Company (LILCO), which is readily understood from Figure 3.21.[28] On 11 April 1995, Timothy J. Driscoll, Director, Research and Development, LILCO, informed me this project had been canceled due to problems encountered with the builder (a third party).

The paragraphs above outline some of the American progress in petro-electric automobiles. What was happening in international markets is the subject of Chapter 4.

# Note 1—Additional Comments

After reading this chapter, Professor Philip T. Krein, who was previously cited, wrote the following to me on 7 December 1994 :

> When the a-c drives were first attempted late in the 1960s (to which you were a major contributor), their advantages quickly became clear. Today, it is possible to purchase high-performance a-c drives with peak ratings beyond 150 kW. Almost any passenger car can be configured to match stock vehicle performance with drives such as these. Of course, the hybrid's engine system must be properly sized as well. The drives I refer to are off-the-shelf industrial products, not custom laboratory prototypes. Combined with conventional squirrel-cage motors, such drives provide almost double the power-to-weight ratio of high-performance spark-ignition engines.
>
> It is probably true that hybrids do not offer dramatic fuel economy improvements by themselves. After all, the purpose of hybrids is to gain the energy storage advantages of liquid fuel. Highway economy, for example, will not change much for a hybrid. Series designs show some improvement in urban driving, if the engine efficiency is kept high. Regenerative braking potentially recovers about 20% in the energy used in the Federal Urban Driving Cycle. Emissions are the most significant aspect of hybrid operation. In a series configuration, the engine can be decoupled from performance needs. This (mode) allows the engine to operate at minimum BASF whenever it runs. Emission can be reduced in at least five ways:
>
> 1. Operate at minimum BASF to minimize tailpipe exhaust per unit of energy input.
>
> 2. Engine transients can be avoided. Transients are thought to account for a substantial fraction of emission.

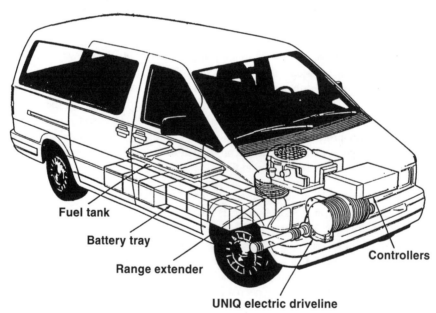

*Figure 3.20. The Unique Mobility Inc. drive system. Top photo shows the van.
(See Ref. 27. Unique Mobility Inc.)*

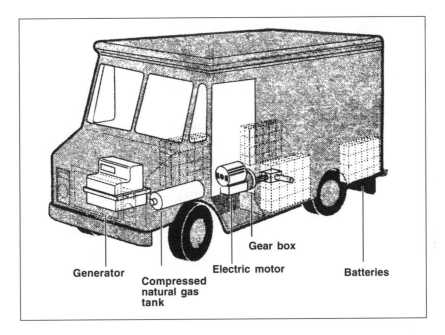

*Figure 3.21. The Grumman-LILCO natural-gas battery van. (Northrop-Grumman)*

3. The catalyst and exhaust treatment systems can be designed optimally for the pre-determined engine operating point to provide the best possible performance.

4. Engine starts can be anticipated without influencing vehicle operation. This permits the straightforward use of catalyst preheaters to reduce cold-start effects.

5. There are no emissions associated with idling conditions. The engine must operate only when its output can provide useful work.

These (gains) are in addition to the advantages of a smaller engine and to the possible use of pure electric modes for short-range driving. From this discussion, we conclude that it is possible to match conventional vehicle performance with both parallel and series designs. The most costly added subsystem—the a-c drive electronics—is rapidly becoming cheaper. With a series design, tailpipe emission reductions of 80% or more compared to 1994 levels are quite plausible. Our laboratory data for the University of Illinois hybrid car suggests that 90% reductions can be achieved with a series hybrid running alcohol fuels. We believe that the federal ultra-low emission vehicle (ULEV) targets serve as a realistic benchmark for hybrid system designs. Parallel designs appear to offer more modest emission reductions and no longer have a performance edge over series architectures.

## Note 2—A Petro-Electric Bus

An elegant, low-floor, petro-electric city bus (Figure 3.22) has been developed, the details of which should be referred to in the original article.[29] This work was performed on the FTA/ New York State Consortium Electric Bus Development Program, which managed the project. The drive system is in series. A constant-powered, low-emitting, Cummins diesel engine drives a 100-kW Onan alternator. This power source charges a 270-cell, 25-kWh capacity, nickel-cadmium battery, the output of which powers four independent inverters providing variable frequency power to a General Electric, 75-kW induction motor at each of four rear drive-wheels of the bus. Figure 3.23 illustrates the drive system. The control-motor module was taken from the microprocessor-based U.S. Department of Energy Modular Electric Vehicle Program, now a commercial product of GE Drive Systems. The bus is provided with regenerative braking. With motors at each driving wheel, conventional axles are not required, allowing for a low-floor, easy-access vehicle.

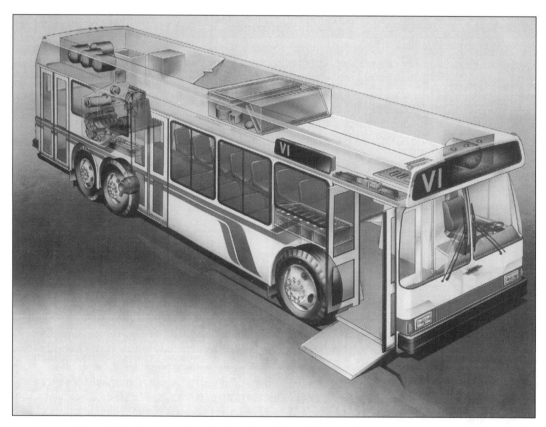

*Figure 3.22. A petro-electric, low-floor bus with an induction motor at each of four drive wheels. (General Electric Co.)*

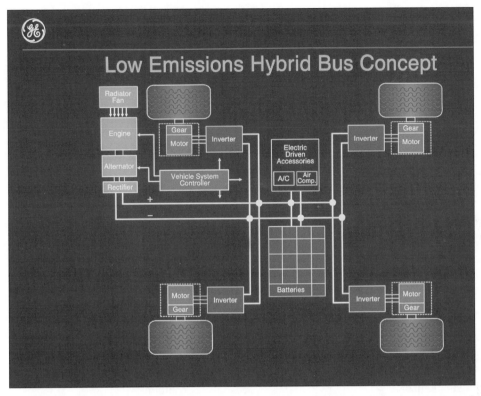

*Figure 3.23. The bus is a series hybrid electric system with an a-c induction motor and reduction gear integrated into each of its four drive wheels. (General Electric Co.)*

# Note 3—The Briggs & Stratton Hybrid Electric Vehicle

For the interested reader, the following provides detailed specifications for the Briggs & Stratton hybrid electric car:

**CHASSIS**

Front Suspension:

Unequal A-arms, coil springs

Rear Suspension:

Solid axle-differential, leaf springs

**POWERPLANTS**

Engine: Briggs & Stratton Model 42

18 hp @ 3600 rpm

2 cylinder, 4 cycle

## CHASSIS *(continued)*

Battery Pack:

Solid axle, leaf springs

Steering:

Rack and pinion, collapsible column

Wheels:

Aluminum alloy 13 × 5.5

Tires:

Goodyear Arriva Radials

P165/80R13

Brakes:

Regenerative electric plus standard
hydraulic

Front—9-inch diameter discs

Rear—9-inch diameter drums

## BODY DIMENSIONS

Height: 50 inches

Width: 64 inches

Length: 174 inches

Wheelbase: 86/112 inches

Weight: 3200 lb (1200 lb front, 1000 lb
each rear axle)

## INTERIOR DIMENSIONS

Front leg room: 46-1/2 inches to seat
back

Front head room: 35-3/4 inches

Rear leg room: 25-3/4 inches

Rear head room: 26 inches

Passenger capacity: 2 adults, 2 children

## HEATER/DEFROSTER

Espar Products gasoline fueled

8000 BTU per hour

Fuel use: 0.071 gallons per hour

## POWERPLANTS *(continued)*

Air cooled, integral fan

42.33 cubic inches (694 cc)

Electric Motor:

Baldor series-wound d-c

8 hp continuous

20 hp intermittent

## BATTERIES

Traction Pack:

Globe Battery Div. lead-acid

(12) 6 volt (72 volts total)

250 amp-hr

1000 lb total weight with trailer

Accessory:

Globe Battery Div. lead-acid

12 volt deep discharge

68 amp-hr

## CLUTCH, TRANSMISSION

Automatic Clutch:

Borg-Warner Duo-Cam one-way
clutch

Manual Clutch:

9-inch dry disk automotive

Transmission:

4 speed manual

Ratios: 3.98/2.14/1.42/1.00

Differential:

4.11 ratio

# References

1. Renner, Roy A., and O'Connell, Lawrence G., "The Hybrid Vehicle Revisited," Eighth International Electric Vehicle Symposium, Washington, DC, 1986.

2. Bhate, Suresh, Chen, Hain, and Dochat, George, *Advanced Propulsion System Concept for Hybrid Vehicles*, U.S. Department of Energy, 1980.

3. Stephenson, R. Rhoads, *et al.*, "Should We Have a New Engine?" Technical Report, JPL SP 43-17, Jet Propulsion Laboratory, California Institute of Technology, Pasadena, CA, August 1975, Chapter 9, p. 4.

4. Andon, J., and Barnal, I., "Emission Optimization of Heat Engine/Electric Vehicle," Minicars Inc., Report EHS 70-107 to U.S. Environmental Protection Agency Air Pollution Control Office, January 1971.

5. Wouk, Victor, "An Experimental ICE/Battery Electric Hybrid with Low Emission and Low Fuel Consumption Capability," SAE Paper No. 760123, SAE Automotive Engineering Congress, 23 February 1976, Detroit, MI, Society of Automotive Engineers, Warrendale, PA.

6. Wouk, Victor, "Decreasing On-Board Fuel Consumption in Heat Engine/Battery Electric Hybrids by Battery Depletion," SAE Paper No. 780295, 27 February 1978, Society of Automotive Engineers, Warrendale, PA, 1978.

7. Wouk, Victor, "Test Procedures for Hybrids—A Review of Proposals to Date," SAE Paper No. 82069, SAE International Congress, 22–26 February 1982, Society of Automotive Engineers, Warrendale, PA, 1982.

8. Wouk, Victor, "Electric Vehicles Are Practical with Today's Technology," 32nd IEEE Vehicular Conference, San Diego, CA, 23–26 May 1982.

9. Wakefield, Ernest H., "A High Performance A-C Electric Drive for Vehicles," Proceedings of the First International Electric Vehicle Symposium, Phoenix, AZ, 5–7 November 1969. Also Electric Vehicle Council, New York, 1969.

10. Wakefield, Ernest H., *History of the Electric Automobile: Battery-Only Powered Cars*, Society of Automotive Engineers, Warrendale, PA, 1994.

11. Murphy, Gordon J., "Consideration in the Design of Drive Systems for On-the-Road Electric Vehicles," Proceedings of the IEEE, **60**, 12 December 1972, p. 1522.

12. Edwards, Geoff, "Enemies of the IC Engine," *Engineering*, 6 June 1969.

13. "Luxury Electric Car Unveiled in Detroit," *Electric Vehicles,* **61**, 2 June 1975, pp. 4–7.

14. McKee, R.S., personal communication, 24 September 1993.

15. *Electric Vehicle News,* **10,** 4, 1981, p. 33.

16. Wouk, Victor, personal communication, 20 May 1982.

17. "Silver Volt Covers 300 Miles in 12 Hours," *Electric Vehicles,* **66,** 4 December 1980, pp. 12–13.

18. *Electric Light and Power*, July 1980.

19. "Hybrid Prototype Gives New EV Potential," *Electric Vehicle News,* **9,** 1 February 1980, p. 14.

20. Brams, Stanley H., "Electric Cars: Will 'Plug It In' Replace 'Fill It Up'," *Home & Away,* January/February 1981, p. 8–10. Also: Salgado, Robert J., "1980 Briggs & Stratton Hybrid," *Autoweek,* 21 June 1993, p. 48.
21. "Hybrid Takes Shape in Schenectady," *Electric Vehicles,* **66,** 4 December 1981, p. 14.
22. Burke, Andrew F., "The Hybrid Test Vehicle (HTV)—Concept Through Fabrication," SAE Paper No. 820266, SAE International Congress, 22–26 February 1982, Detroit, MI. For a continuation, read "Hybrid Vehicle Program: Final Report," Contract No. 955190; submitted to Jet Propulsion Laboratory, California Institute of Technology, Pasadena, CA; submitted by General Electric Company, Corporate Research and Development, Schenectady, NY.
23. Sandberg, Joel J., "Hybrid Vehicles—Costs and Potential," SAE Paper No. 810271 (P-91), Society of Automotive Engineers, Warrendale, PA, 1981.
24. McInnis, Stirling A., "An Electric Hybrid Car," International Electric Vehicle Symposium, Washington, DC, 1986.
25. Wipke, Keith, personal communication, 12 October 1993.
26. Larsen, Robert P., *Innovations in Design: 1993 Ford Hybrid Electric Vehicle Challenge*, SP-980, Society of Automotive Engineers, Warrendale, PA, 1994.
27. Nelson, T., Lighthipe, R., and Cambrier, C., "A Natural Gas Hybrid Minivan," Tenth International Electric Vehicle Symposium, Hong Kong, 3–5 December 1990, pp. 777–783.
28. Wald, Matthew L., "More Join the Quest for Cleaner Cars," *The New York Times*, 16 December 1993, p. C4.
29. King, R.D., Haefner, K.B., Sallasoo, L., and Koegl, R.A., "Hybrid Electric Transit Bus: Pollutes Less, Conserves Fuel," *IEEE Spectrum,* July 1995, pp. 26–31.

# CHAPTER 4

# Modern International Petro-Electric Automobiles

The use of hybrid electric vehicles, as shown in the previous chapter, knows few boundaries. Although the battery-only electric car is attractive in efficiency, in quietness, and in freedom from pollution, its range is usually perceived as being too limited. As a result, a viable petro-electric car is being sought abroad and in the United States.

## British Petro-Electric Automobiles

In 1977, personnel connected with Queen Mary College of London University were developing a diesel-electric vehicle drive. The power system was a diesel engine which could be coupled to the wheels and generator. Electro-magnetic clutches could be incorporated between the engine, the wheels, and the generator. Another system to be investigated uses the engine and motor alone. In this case, the motor operates in the regenerative mode in slowing the vehicle or in idling, and the battery is charged at such times. At any chosen period (e.g., before entering the downtown area), the engine could be switched off and the vehicle could be operated as an electric vehicle. This work again shows the effort to overcome the range limitation of the battery-only electric automobile.

Again in Britain, Lee-Dickens Ltd. of Desborough, Northampton, has demonstrated a petro-electric car, as shown in Figure 4.1. A 60-lb (27-kg), fixed-speed petro-engine feeds power to a 120-lb (54-kg) battery, which powers an electric motor. The latter is equipped with regenerative

*Figure. 4.1. The Lee-Dickens Ltd. automobile series petro-electric car.* (Electric Vehicles)

braking to recharge the batteries.[1] The range of the vehicle is limited only by the quantity of gasoline carried.[2] This system is seen as a series system. Presumably, the engine might be turned off for inner-city driving. Another innovator working on petro-electric automobiles is Professor A.E. Corbett of Warwick University. His approach uses disc-type motors,[3] one of which is illustrated in Chapter 24 of *History of the Electric Automobile: Battery-Only Powered Cars.*[4] William H. Shafer, retired spokesman for Commonwealth Edison Company, believes that a petro-electric car in daily use will save approximately half the normal liquid fuel, the balance being electricity derived from coal, natural gas, wind, hydropower, or nuclear power.[5]

## Canadian Petro-Electric Automobiles

In the Clean Air Race of 1970, the University of Toronto entered a petro-electric vehicle dubbed *Miss Purity* (Figures 4.2 and 4.3), designed and built by a staff-student committee headed by Professor F.C. Hooper of the Mechanical Engineering Department. (See Note 1 at the end of this chapter for a list of others involved in the project.) This nearly all-Canadian conceived and contributed automobile has a polyester/fiberglass body and bears gull-wing doors. "It was," Hooper writes, "(1) basically a propane-fueled internal combustion engine drive, but can operate (2) all-electric on batteries, (3) as a hybrid, and (4) on combined propane/battery power. The vehicle can be readily converted to gasoline for future studies."[6] With the front end a 1970 Chevrolet *Chevelle*, midsection and rear end being University built, bearing aluminum wheels, and Dunlop 185/15 radial ply tires, the vehicle is powered both by a 302-in.[3], V-8 engine and a 16-hp, d-c shunt motor, 100 volts, 120 amps with 2:1 timing belt drive to the transaxle input. The motor has constant torque at 0 to 3700 rpm and constant horsepower from 3700 to 8000 rpm. A solid-state, fixed-frequency, variable-pulse-width chopper controls the motor. Speed control is by the field circuit.

*Figure 4.2.* Miss Purity *from the University of Toronto.* *(F.C. Hooper)*

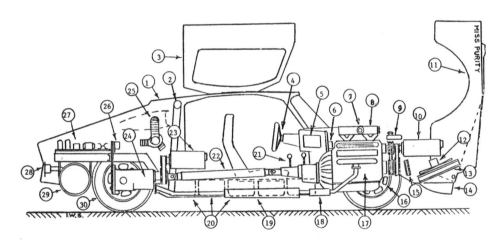

| | | |
|---|---|---|
| 1. FIBERGLASS BOBY | 11. FOREBODY (RAISED) | 21. STICK SHIFTS |
| 2. SAFETY ROLL BAR | 12. ELECTRIC FANS | 22. ENGINE DRIVESHAFT |
| 3. GULL-WING DOOR | 13. RADIATOR | 23. D-C MOTOR AND DRIVE |
| 4. SAFETY STEERING WHEEL | 14. AIR SCOOP | 24. CORVAIR TRANSAXLE |
| 5. ELECTRIC LOGIC CIRCUITS | 15. OIL COOLER | 25. EQUIPMENT FANS |
| 6. EXPANSION VALVE | 16. ALUMINUM WHEELS | 26. AMP-HR METER |
| 7. CARBURETOR | 17. V-8 ENGINE | 27. CHOPPER CONTROL |
| 8. SECONDARY PLENUM | 18. CATALYTIC MUFFLER | 28. PROPANE FILLER |
| 9. RADIATOR HEADER TANK | 19. MAIN MUFFLER | 29. PROPANE TANK |
| 10. D-C GENERATOR AND DRIVE | 20. BATTERIES | 30. RADIAL PLY TIRES |

*Figure 4.3. Identified elements for* Miss Purity. *(F.C. Hooper)*

The ten *Exide* lead-acid batteries (10 kWh) are charged from a 16-hp, 140-volt, 90-amp, d-c generator with a 2:1 timing belt drive to the front of the engine. An additional 12-volt battery supplies ancillary power.

# French Petro-Electric Vehicles

With such a rich history in the design of petro-electric vehicles, as outlined in Chapter 2, it is only natural that French engineering would continue to exploit a combination of energy sources to overcome the range limitation of battery-only powered automobiles. For this continuation, *Laboratoire et d'Architecture d'Analyses des Systèmes du CNRS* of Toulouse, France, has been developing the controls for a petro-electric system in which the thermal engine provides steady-state power to the wheels while the battery-powered electric motor absorbs the fluctuation of power demand.[7] In addition, it is reported that the electric motor-generator reclaims for the battery 20 to 25% of the kinetic energy, the vehicle being equipped with regenerative braking. Moreover, with the thermal engine operating at constant load, the constant power from the engine drives the generator to charge the battery. Similar to Dr. Charles Proteus Steinmetz's path to his revolutionary electric vehicles as described in Chapter 5 of *History of the Electric Automobile: Battery-Only Powered Cars*,[4] the approach of the Laboratoire has had a mathematical base. Figure 4.4 illustrates this successful solution.

# German Petro-Electric Automobiles

Robert Bosch GmbH of Stuttgart has tested a parallel dual-powered sedan with a total drive system weighing only 880 lb (399 kg). With the clutch engaged, the engine, electric motor, and universal shaft rotate at the same speed. The engine delivers power directly to the differential, with a power loss of approximately 10%. At low speeds, the gasoline-powered engine may be disconnected from the electric motor, and then the car can proceed on battery power. The motor is equipped for regenerative braking.[8]

# A Robert Bosch Hybrid Electric Automobile

Engineers from all industrial countries are striving to overcome the perceived short range of the battery-only powered electric car, as seen from the writings of R. Rhoades Stephenson *et al.,* who write:[9]

> Two systems of particular interest have been developed in the Federal Republic of Germany within the past year (1974). The Robert Bosch system is an on-off battery/heat engine hybrid in which the engine is shut off below a speed of 20 mph (32 km/h).[10] The system has been installed and tested in a Ford *Escort* station wagon. The 44-hp conventional Otto engine and 21.5-hp electric motor-generator are mounted in-line but are separated by an automatically operated dry friction clutch with manual override. The electric motor takes the position formerly occupied by the transmission, which is not used in the

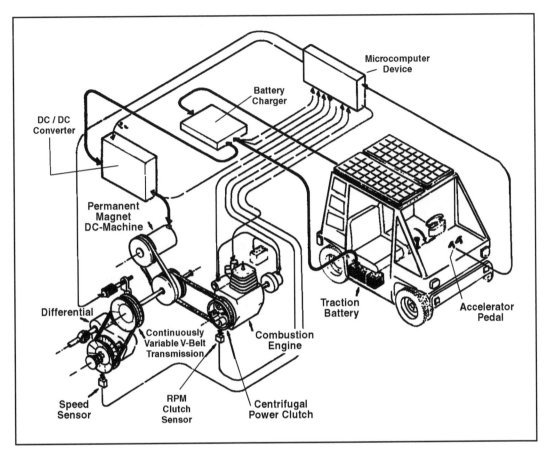

*Figure 4.4. A French petro-electric system. (Ref. 7. Professor J.C. Marpinard, Centre National de la Recherche Scientifique, Toulouse, France)*

hybrid modification. A manually operated jaw-type clutch couples the electric motor-generator output to the propeller shaft. To compensate for the loss of low-speed torque resulting from the removal of the transmission, the rear axle ratio has been changed from 4.11:1 to 5.83:1, which decreases the maximum speed from 80 to 68 mph (129 to 109 km/h). (Approximately) 445 lb (200 kg) of nickel-cadmium batteries are recessed into the floor pan on each side of the drive tunnel ahead of the rear axle. The total conversion adds 882 lb (400 kg) to the overall vehicle weight.

In urban driving, the car may be operated on battery power alone or in the hybrid mode. In the latter case, the engine starts automatically at 20 mph (32 km/h) vehicle speed and stops automatically when the speed falls below 15 mph (24 km/h). Battery recharging while in this mode can be adjusted during intermittent periods of engine operation to maintain a nominally constant battery charge state, or at least reduce the rate of battery discharge.

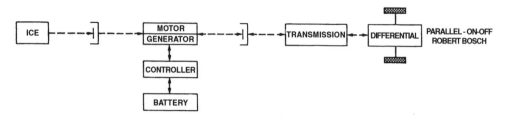

*Figure 4.5. Drive train of the Bosch hybrid electric motor car. (Jet Propulsion Laboratory)*

During constant-speed driving, the battery is recharged by operating the motor-generator as a generator and absorbing excess engine power. If the battery becomes discharged due to sustained operation in congested traffic, the car must be stopped, the propeller shaft jaw clutch disengaged, and the battery recharged by the engine through the motor-generator. The battery has sufficient storage capacity to operate the car over a range of 13 miles (21 km) in stop-and-go uphill and downhill driving without engine assist.

The 67-in.$^3$, four-cylinder engine is equipped with a 'low-emission' carburetor and exhaust thermal reactor. Urban driving emissions in the hybrid mode are 2.83 g/mi. for HC, 5.52 g/mi. for CO, and 3.67 g/mi. for $NO_x$. With further engine modifications, improvements in emission performance to 2.5/5.5/2.0 g/mi. are expected to be possible.

Figure 4.5 shows the basic blocks of this system.

## The Volkswagen-Electricité Neuchâteloise S.A. System

In the mid-1980s, Volkswagen AG and Electricité Neuchâteloise S.A. together designed and placed in a 1982 *Golf* sedan a 44-kW petrol engine and a 5-kW electric motor.[11] With the former power plant alone, the maximum speed on the level was 148 km/h (92 mph). With battery only, the maximum speed was 52 km/h (32 mph). The engine and motor would not operate simultaneously. As previously mentioned, the French chose to operate the spark-ignition engine at a constant speed during all driving, with the electric motor providing the transient power needs. However, the Germans reversed the case: in city driving, the electric motor would furnish routine propulsion at speeds commensurate with city driving, while the petrol engine accommodated the higher power demands.

Using the German city of Brunswick with a population of 200,000 (as well as the European test-driving program) as an extensive but reproducible test track (the same city streets) between the hours of 9:00 A.M. and 6:00 P.M., the share of the combustion engine of the total distance was 17.6%, while that of the electric motor was 56.1%. Coasting of the vehicle made up the balance. It is stated that these results are very close to the figures obtained from the European test-driving cycle. The 5 kWh of battery capacity (300 lb, or 136 kg) gave a range of 30 miles (48 km) in the petro-electric mode of driving, 166 watt-hours/mile. In terms of emissions, using the European cycle-test, the dual-powered *Golf* consumed 33% as much liquid fuel consumption as the test-standard spark-ignition *Golf*. The dual-powered *Golf* also showed carbon dioxide decreased by 79%, hydrocarbons by 57%, and nitrogen oxides ($NO_x$) by 20%. The $NO_x$ decreased less because "the combustion engine covers only the acceleration phases (when) higher levels of $NO_x$ are produced than is the case during phases involving lower output requirements." For holiday highway driving, the German authors recommended the battery be removed to avoid carrying the extra weight. I have found that removing batteries can be a dirty and hazardous job. Thus, you might conclude something is lacking in such a solution.

For completeness on German petro-electric vehicles, Figures 4.6 and 4.7 show two German examples. Figure 4.6 demonstrates the all-electric battery trailer-type interchange. Figure 4.7 shows a vehicle with an all-electric drive system in urban areas and a series petro-electric system elsewhere. In the former, 20 of these vehicles were built and operated from 1976 to 1980 in Dusseldorf. The system was technically feasible but was operationally inconvenient and uneconomical. As Dr. Victor Wouk states, "The bus had to leave its route for battery exchange two or three times a day. In addition, the capital cost of the battery exchange equipment and the additional capital cost of batteries under charge made the system uneconomical." In 1980, 12 of the vehicles were revised so "at the bus-line terminal, the (vehicle) pole system is raised, making contacts for battery-charging, vehicle grounding, etc."[12]

*Figure 4.6. Battery exchange system for electric buses in Germany. (Victor Wouk)*

*Figure 4.7. Petro-electric bus used in Germany: all-electric in urban areas, and series petro-electric elsewhere. (Victor Wouk)*

# Opel *Twin* Petro-Assist Automobile

Opel has built a novel type of hybrid electric vehicle. David Scott writes[13]:

> Want an electric car for nonpolluting trips around town, but also want a gas-engined one with go-anywhere range for longer cross-country drives? Opel offers both in its *Twin* concept car, which takes interchangeable rear-end power modules with plug-in connections. (See Figure 4.8). Fuel economy and environmental benefits are the aim of this novel dual-mode concept. Today you've got a 250-mile turnpike run, so you drop in at your dealer. He exchanges your electric-power module for a gasoline one in a matter of minutes on a rental basis and then swaps them back again when you return from your trip.
>
> Average fuel consumption with the gas module is said to be 67 mpg (28 km/l) in a four-seater car capable of 87 mph (140 km/h); it accelerates to 62 mph (100 km/h) in 20 seconds. That's from an 800-cc, three-cylinder, 34-hp engine with a six-speed automatic transmission programmed to minimize (fuel) consumption at all speeds. The package includes the rear axle, wheels, and gas tank, and quick-release connections for all driver controls. Also completely self-contained, the alternative electric drive train has two wheel-hub motors that dispense with the usual differential and half-shafts. Each of these three-phase asynchronous (induction motor) units develops 14 hp and accelerates the *Twin* to 30 mph (50 km/h) in seven seconds with a 75-mph (121-km/h) top speed. The batteries are an advanced high-energy 'cold' type with a lithium-carbon base in a 550-lb (250-kg) pack that has a 155-mile (250-km) range in city traffic and a recharge time of just two hours.

Table 4.1 describes some characteristics of the Opel *Twin*.

*Figure 4.8. The Opel* Twin *with two interchangeable power plants. First shown in 1992 in Geneva. (Adam Opel, Germany)*

**TABLE 4.1**
**CHARACTERISTICS OF THE OPEL *TWIN***
**(TARGET VALUES)**

|  | *Twin* with Gasoline Engine | *Twin* with Electric Power |
|---|---|---|
| **Engine Data:** | | |
| Output (kW/hp) | 25/34 | 2 × 10/14 |
| at Engine/Motor Speed (rpm) | 5,500 | 4 – 10,000 |
| Maximum Torque (Nm) | 72 | 2 × 24 |
| at Engine/Motor Speed (rpm) | 2,500 | 0 – 4,000 |
| | | |
| **Transmission:** | | |
| Gearbox | Six-Speed | Fixed Ratio |
| Final Drive Ratio | 3.74 | 7.0 |
| | | |
| **Dimensions and Weights:** | | |
| Length/Width/Height (mm) | 3,475/1,630/1,490 | |
| Wheelbase (mm) | 2,335 | |
| Luggage Compartment Volume (1) | 300/500 | |
| Tire Size | 125/80 R17 | |
| Unladen Weight/Payload (kg) | 540/390 | 740/350 |
| Fuel Tank (1) | 20 | — |
| Battery Capacity (kWh) | — | 29 |
| Refueling/Recharging Time (min) | 2 | 120 |

**TABLE 4.1 *(continued)***

|  | *Twin* with Gasoline Engine | *Twin* with Electric Power |
|---|---|---|
| **Performance Data:** | | |
| Maximum Speed (km/h) | 140 | 120 |
| Acceleration | | |
|    0–100 km/h (s) | 20 | 30 |
|    0–50 km/h (s) | 5.5 | 7 |
| Fuel/Energy Consumption per 100 km | | |
|    At 90 km/h (1 or kWh) | 2.9 | 11.5 |
|    In Urban Traffic (1 or kWh) | 3.7 | 11.5 |
|    Average Value (1 or kWh) | 3.5 | 11.5 |

# European Ford Petro-Electric Car

The European Ford petro-electric motorcar complements the battery-only American Ford *Ecostar* illustrated and discussed in Chapter 19 of *History of the Electric Automobile: Battery-Only Powered Cars*.[4]

David Scott writes[14] that the Ford petro-electric motorcar:

> ...based on the *Escort* wagon, has an orbital two-stroke, three-cylinder engine that produces (68) horsepower from only 1.2 liters. It matches a conventional 1.6-liter, four-stroke engine in output and flexibility but is a third lighter and more compact, reducing vehicle weight by 65 lb (30 kg). For city driving, there's a (30 kW) water-cooled, asynchronous (squirrel-cage, induction) electric motor whose small (7 kWh) nickel-cadmium battery pack provides a 30-mile (50-km) range. In this electric mode, acceleration to 30 mph (50 km/h) is a competitive 6.8 seconds, and a top speed of 60 mph (97 km/h) is possible. At 40 mph (64 km/h), however, the gas engine starts automatically and takes over driving both the car and the electric motor, which then functions as a generator to charge the batteries. Top speed in this mode is 95 mph (153 km/h), with acceleration to 60 mph (97 km/h) at a leisurely 22 seconds.

This hybrid electric car is considered to be of parallel type, as illustrated in Figure 2.1 of Chapter 2.

Scott continues:

> But when you need an extra surge—to pass another car, for instance—this car parts company with other hybrids. At full acceleration, the electric motor becomes active again and joins the gasoline engine in driving the vehicle. The combined power of 138 hp and 148 lb-ft of torque (a 66% boost) cuts the 0-to-60-mph acceleration time to a sporty 10.6 seconds. There's no gain in top speed. With this tandem power, both drive units can be kept relatively small. The two are coupled via a special transfer case to the car's

four-speed automatic transmission. An international team of Ford engineers from Germany, Britain, and the United States developed the car in cooperation with the Institute of Automotive Engineering in Aachen, Germany.

Figure 4.9 shows a cutaway version of this motorcar.

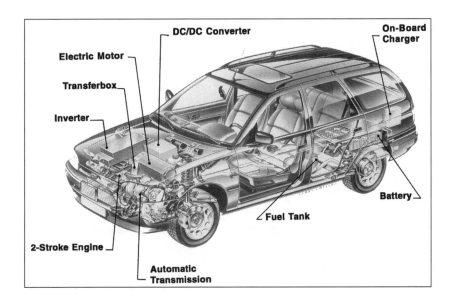

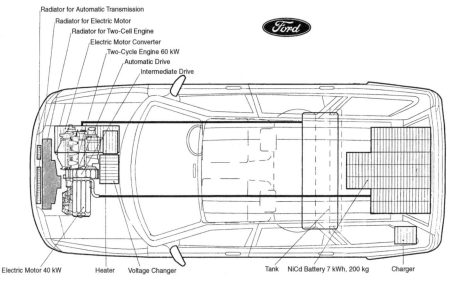

*Figure 4.9. The European Ford hybrid electric car,* Escort Turnier, hybrid Antrieb *(top), with identifiable elements (below). (Ford Werke, Köln)*

# The Volkswagen Petro-Electric Car—The *Chico*

Shown in Figure 4.10 and seen at the January 1994 Detroit automotive show was the Volkswagen petro-electric entry.[15] The vehicle, which was built as a demonstration project, is equipped with a 72-volt, 30-amp/hr nickel-metal hydride battery powering a three-phase induction motor, to be used for low-speed, nonpolluting, city driving. For quick acceleration and for speeds greater than 35 mph (56 km/h), a gasoline engine complements the electric motor. The car is seen as the parallel type.

**Table 2: Vehicle Data**

| | |
|---|---|
| Aerodynamic drag coefficient | 0.34 |
| Front area (m²) | 1.9 |
| Rolling resistance | 0.009 |
| Curb weight (kg) | 785 |
| Payload (kg) | 350 |
| Seats | 2+2 |
| Luggage capacity (l) | 60/600 |
| Length (mm) | 3150 |
| Width (mm) | 1600 |
| Height (mm) | 1480 |
| Wheelbase (mm) | 2050 |
| Headroom, front (mm) | 1000 |
| Leg room, front (mm) | 1132 |
| Liftover height (mm) | 682 |
| Shoulder room, front (mm) | 1290 |
| Tires | 155/65 R 14 |

**Table 3: Engine Data**

| | |
|---|---|
| **IC engine** | Two-cylinder four-stroke gasoline electronic injection, catalytic converter |
| Capacity (cm³) | 636 |
| Compression | 9.3:1 |
| Power (kW @ rpm) | 25 @ 6000 |
| Max. torque (nm @ rpm) | 45 @ 3000 |
| Fuel tank (l) | 20 |
| Transmission | Five-speed semi-automatic gearbox |
| **Electric motor** | Three-phase asynchronous induction motor |
| Power cont./max. (kw) | 7/9 |
| **Traction battery** | NiMeH 72 V/30 Ah |

**Table 4: Driving Performance and Energy Consumption**

| | |
|---|---|
| Top speed (km/h) | 131 |
| Acceleration time | |
|   0–100 km/h (s) | 30 |
|   0–80 km/h (s) | 19 |
| Energy consumption | |
|   ECE cycle [(1 + kWh)/100 km] | 1.41 + 13.0 |
|   90 km/h const. (1/100 km) | 4.3 |
|   120 km/h const. (1/100 km) | 6.4 |

*Figure 4.10. The Volkswagen petro-electric car—the* Chico.
*(Volkswagen, V. Platzer and Dr. S. Brüdgam)*

# Japanese Petro-Electric Vehicles

The Toyota Company of Japan, having developed a battery-only powered van which was demonstrated in 1967 (see Chapter 23 of *History of the Electric Automobile: Battery-Only Powered Cars*[4]), announced a series type of petro-electric automobile in 1976. A gasoline engine-generator continually charged a battery system which in turn powered two d-c electric motors. Each motor independently drove a front wheel, an approach first used by Krieger in France as illustrated in Figure 7.13 of Ref. 4. This dual motor system has the obvious advantages of smaller armature currents, easier current control, and lighter and smaller motors, characteristics which more than compensate for the additional electronics. Top speed of this designated GT-54 was given as 72 mph (116 km/hr).[16] In late 1997, Toyota introduced a parallel-type sedan, the *Prius*, as a production vehicle for its domestic market. This vehicle was the first production hybrid vehicle offered by a major car manufacturer in what many saw as a "green" revolution caused by slowly rising gasoline prices and increasing urban air pollution.

The *Prius* features a 1.5-1 gasoline engine, nickel-metal hydride batteries, and regenerative braking. As for its motor, Jack Yamaguchi reports[17]:

> The electric motor/regenerator is a permanent magnet, synchronous, ac, watercooled type rated at 30 kW at 940–2000 rpm and 305 N/m peak torque produced at 0–940 rpm. The parallel hybrid system employs a separate generator, which supplies power to the propulsion motor that recharges the battery, and by modulating the amount of electricity generated, controls the planetary-gear-type transaxle's continuously variable transmission function. Both the motor/regenerator and the generator are of Toyota's own design and manufacture.

The *Prius* reportedly achieved 66 mpg (28 km/l) in road tests. The high initial cost of the vehicle (estimated to be $41,000) was not reflected in the price to the consumer (2.15 million yen, or approximately $20,000), resulting in an initial loss for Toyota of more than $20,000 per car sold. In the first month of production, orders were received for 3,500 vehicles.[18] Toyota expects high production levels to reduce costs, eventuating in a profit. Ronald M. Lehotsky of Pittsburgh, Pennsylvania, forwarded a report on a 1970 Mazda-built battery-powered city car as shown in Figure 4.11.[19] Later models are reported also to contain a rotary internal combustion engine to keep the batteries charged.

# Swedish Petro-Electric Automobiles

Swedish engineers had attacked the range limitation of battery-only powered electric vehicles by developing a hydraulic system for regaining the kinetic energy of movement when a halt is demanded. In a sense, a mechanical system was to take the place of electrical regenerative braking. This technique was reported in Chapter 23 of Ref. 4. The City Council of Los Angeles, California, which is allegedly the most air-polluted city in America, was frustrated by what it perceived as the slow development of commercial electric vehicles by the American Big Three motorcar manufacturers and took action. The Council advertised for bids for

*Figure 4.11. A Mazda-built petro-electric motorcar. Specifications are as follows: length is 91.7 in. (2330 mm); width is 57.1 in. (1450 mm); height is 62.6 in. (1590 mm); weight is 991 lb (450 kg); engine is a 200-cc rotary engine; and maximum speed is 25 mph (40 km/h). (Mazda (North America), Inc., Masashi Aihaira)*

an electric vehicle that could adequately service at expressway speeds the large area of Los Angeles. The winner of the solicitation was the Clean Air International Automotive Design Ltd. of Sweden, also known as Clean Air Transport. The instrument designated by the Council to execute the contract was The Department of Water and Power (DWP), a historically innovative city-owned organization. With the backing of the DWP, Clean Air Transport plans to introduce 1,000 two-door hatchbacks, designated *LA 301*, in 1993 and 30,000 by 1995.[20] Detailed engineering of the vehicle was done by International Automotive Design based in Sussex, England. ("An HEV [Hybrid Electric Vehicle] that works! The problem is in financing." Victor Wouk, 1994)

Electric energy for the *LA 301* is from 216-volt sealed lead-acid batteries located under the car's center floor. A 57-hp series-wound d-c electric motor drives the front wheels through an automatic transmission.[21] Battery capacity is a reported 18 kWh, and a probable weight is 1080 lb (490 kg). An onboard charger capable of service from 110/220-volt outlets is provided. With a stated range of 60 miles (97 km), the condition of driving being omitted, energy consumption would be approximately 300 watt-hours/mile (180 watt-hours/km). Gasoline from a seven-gallon gas tank fuels an auxiliary engine with a reported ultra-low emission. This power source, a 25-kW (33 hp), four-cylinder, 650-cc water-cooled, internal combustion engine with fuel injection, fitted with a preheated catalytic converter, is intended to augment electric power at speeds greater than 30 mph (50 km/h), according to John Samuels, director of engineering for Clean Air Transport. The internal combustion engine source is Japan.

According to Daniel B. Wood, when driving:

> ...for journeys up to 150 miles (250 km), the driver selects XR (extended range), which allows the auxiliary engine to cut-in at cruising rates greater than 30 mph (50 km/h), (a speed regulated) by a computer.

Such tandem use will allow the battery to be discharged more slowly.

> With a seven-gallon tank, it will also extend the car's range to 150 miles (250 km) and still meet California's strict guidelines for ultra-low-emission vehicles.

> In a third mode, RR (remote recharging), the gasoline engine can recharge batteries while the car is parked away from (an electrical outlet). The computer controls power supplies to both motor (and engine), so there is no direct connection to the accelerator pedal. The car's two-speed gearbox regulates torque while starting and cruising, making the ride far smoother than (that of) conventional cars.

With seats for four adults, two-speed automatic transmission, and conventional controls, dashboard and interior, the car is expected to cost approximately $25,000. Figures 4.12 and 4.13 show the exterior and interior, respectively, of the *LA 301*.

*Figure 4.12. The* LA 301 *hatchback. (Clean Air Transport)*

*Figure 4.13. Interior of the* LA 301 *hatchback. (Clean Air Transport)*

## Petro-Electric Cars of the Former Soviet Union

The Russians' interest in electric drives for personal transportation dates back to Count Moritz Jacobi, a German in the St. Petersburg of Czar Nicholas I in 1838.[22] On the city's canals, which serve as connections to the Neva River, Jacobi propelled a paddle-wheel boat with an electric motor. However, the former USSR's first reported electric car, battery-only powered, was the *Electromobile* of 1968, cited in Chapter 23 of *History of the Electric Automobile: Battery-Only Powered Cars*.[4] (Reference 4 also contains reports on many more Russian battery-only powered electric cars.) With this background, it is natural for Professor Sergei Bannikov of the Moscow Motor Road Institute to relate the development of a "promising hybrid." The vehicle is described as a sedan bearing a constant-speed, gasoline engine driving a generator that serves to charge a set of batteries. This drive system is seen to describe a series type of hybrid electric vehicle. The battery in turn powers the electric motor supplying torque to the wheels. In regular driving, pollution is said to be reduced by 70 to 80%.[23] In later periods, Wouk reports that although the USSR delegation attended the 1970 International Electrotechnical Commission meeting in Stockholm, no representatives to electrical vehicle symposia have appeared since that time.[24]

## A Swiss Petro-Electric Motorcar

One facet encouraging the application of electric drives in Switzerland has been the substantial publicity accorded the Swiss *Tour de Sol* races in that mountainous country, a subject omitted here for space reasons. According to Urs Muntwyler, Commissioner of the Swiss Federal Office of Energy for the promotion program Lightweight Electric Vehicles, headquartered in Zollikofen, Switzerland, these races launched "hundreds of highly efficient racing cars, electric scooters, bikes, trikes, and electric cars... Highly efficient in this case means an energy consumption between 6 and 10 kWh/100 km (100 to 160 watt-hours/mile [The GM *Impact* is 113 watt-hours/mile at 55 mph]), corresponding to a fuel consumption between 0.6 and 1 liter of gasoline per 100 km."

Facing the same range limitation of all battery-only powered vehicles, many Swiss companies have opted for hybrid electric automobiles. One of these companies is Esoro AG of Zürich, Switzerland. Initiating its electric vehicle work in 1985, Esoro has brought forth the Esoro *E301 Coupé* and the *H301 Family*. Here, the letter E represents battery-only; the letter H represents hybrid. Figures 4.14 and 4.15 represent the latter vehicle and its drive system, respectively. Table 4.2 gives the characteristics of both cars.

*Figure 4.14. The Esoro H301 Family hybrid electric car.
(Dr. S.K. Muntwyler, Ing. Muntwyler, Bern, Switzerland)*

*Figure 4.15. The Esoro two-stroke internal combustion engine and a-c induction motor parallel
drive system. (Dr. S.K. Muntwyler, Ing. Muntwyler, Bern, Switzerland)*

## TABLE 4.2
## TECHNICAL SPECIFICATIONS
## ESORO E301 AND H301 FAMILY VEHICLES

|  | E301 Coupé | H301 Family |  |
|---|---|---|---|
| **Dimensions:** | | | |
| Length/Width/Height | 3.07/1.55/1.42 | 3.07/1.55/1.42 | m |
| Curb Weight | 620 | 650 | kg |
| Payload | 380 | 350 | kg |
| Gross Weight | 1000 | 1000 | kg |
| **Body:** | | | |
| Seats | 2 + 2 | 4 | |
| Coefficient of Aerodynamic Drag $(C_w)$* | 0.178 | 1.190 | |
| Frontal Area | 1.8 | 1.8 | $m^2$ |
| Aerodynamic Drag $(C_w \times A)$ | 0.32 | 0.34 | |
| **Drive:** | | | |
| • A-C Induction Motor | | | |
| Max. Torque (0–3000 U/min) | 55 | 55 | Nm |
| Max. Power (3000–8000 U/min) | 21 | 21 | kW |
| Weight (incl. Controller) | 40 | 40 | kg |
| • Two-Stroke Internal Combustion Motor | | | |
| Piston Displacement | | 125 | ccm |
| Power | | 12 | kW |
| Weight (incl. Clutch) | | 13 | kg |
| • Transmission | | | |
| Reduction Ratio | 10:1 | 8:1 | |
| Weight | 11 | 11 | kg |
| **Battery:** | | | |
| System Voltage | 136 | 136 | V |
| Capacity | 9.1 | 9.1 | kWh |
| Weight (incl. Management, etc.) | 230 | 230 | kg |
| **Performance:** | | | |
| Top Speed (E and H) | 120 | 120 | km/h |
| Acceleration 0–50 km/h | 7.5 | 7.5 | s |
| Range (Electric) | 100–150 | 100–120 | km |
| Range (Hybrid) | | 600 | km |
| Power Consumption (Electric) | 10 | 10 | kWh/100 km |
| Power Consumption 90 km/h | | 2.6 | 1/100 km |
| Power Consumption Real Mix | | 6 | kWh and |
| | | 1 | 1/100 km |

* 1/3-scale model measured in the small wind tunnel at Emmen's Eidgenössische Flugzeugwerke February 1994. Subject to change for further improvement. (Esoro AG)

# Note 1—*Miss Purity* Participants

The University of Toronto vehicle was funded largely from bequests by late faculty members, Professors J. Roy Cockburn and William A. Wallace.  Participants in bringing forth *Miss Purity* were as follows:

| | |
|---|---|
| F.C. Hooper, Chairman | I.W. Smith, Co-Chairman |
| P.B. Hughes, Secretary | A.B. Allen, Engine Consultant |
| D.A. Venn, Project Director | D.L. Allen, Vibration Consultant |
| K. Bell, Body Designer | S.D.T. Robertson, Elec. Drive |
| S. Baker, Body Engineer | S. Ng, Elec. Controls |
| R.S. Segworth, Elec. Drive | C.G. Denny, Technical Services |
| J. Otsason, System Eng. | Zs. Krisztics, Technical Services |
| G.E. Shessel, Battery Eng. | F.A. Venn, Coordinator |
| G.R. Tadros, Battery Eng. | L. Bertin, Liaison |

# Note 2—The Otto Engine

Before completing the discussion of petro-electric vehicles and, appropriately in Chapter 4, the section on international vehicles, the history of the Otto engine should be recounted. During modern times, in all but two cases (the GM *Stir-Lec I* and the Ford *Torino*, both of which employed Stirling engines), the prime mover employed for petro-electric drive systems (excluding turbine drives) has been the internal combustion engine perfected by the German, Nikolas August Otto (1832–1891).  Similar to most inventions, the Otto engine was built on the work of many preceding scientists and experimentalists.

Concepts employed in the Otto engine probably had their genesis with the Dutch scientist Christiaan Huygens (1629–1695) who has been primarily remembered for his work in optics, clocks, and mathematics.  However, Huygens considered driving a piston by an explosive charge.  His thoughts turned to gunpowder, but he realized there was an inherent lack of control in such a charge with his *moteur à explosion*.  In 1794, the Englishman Robert Street conceived and built a primitive internal combustion engine. A diagram and an explanation of this engine are available in Ref. 25.  Soon after, Philip Lebon, a Frenchman, conceived of a similar machine in 1799.  At the beginning of the nineteenth century, the vacuum piston engine was developed wherein a spray of cold water was introduced onto the hot combustion gases, thus creating a partial vacuum which in turn caused the piston to move by the higher atmospheric pressure on top of the piston.  Indeed, this principle was not unlike the early steam engine of Thomas Newcomen (1663–1729).

In 1826, Samuel Brown, an Englishman, and two Italians, Eugenio Barsanti and Carlo Mateucci, developed the so-called "free piston engine," operating a rack and pinion drive.  Using the principle of the "open" steam engine with its sliding steam input valves, the Frenchman Jean J. Etienne Lenoir in 1860 developed and manufactured approximately 500 internal combustion engines. G. Smith, a German, showed that compression of the explosive mixture before igni-tion would enhance the efficiency of the Lenoir engine. The ignition used at the time was a heated tube. In 1862, the English theorist Alphonse Beau de Rochas explained the advantages of the four-stroke cycle engine. The sequence of events in this cycle is: 1) intake of the combustible air and fuel mixture with the piston's outward travel, 2) compression of the explo-sive mixture by the return stroke of the piston, 3) ignition and combustion of the charge near the top end of piston travel, 4) expansion of the resulting hot gases during the outward stroke, and 5) expulsion of gases on the following inward stroke. These events are usually described as induction, compression, power, and exhaust strokes. Ref. 26 elegantly illustrates these actions. The four-stroke cycle powers most of today's cars.

Rochas met with Otto, who applied this four-stroke concept to an operating engine, first using natural gas and air as the combustible mixture. In addition, Otto, in the same way as Lenoir, added a flywheel to provide inertia of rotation between the powering strokes, a principle as old as the potter's wheel. Placing the four-stroke engine in production, Otto, before the turn of the century, delivered approximately 35,000 engines. Isaac Asimov, in his scientific dictionary, credits the Otto cycle engine as a prerequisite for the development of both the automobile and the airplane.[27] The hot tube, soon supplanted by the electric spark ignition as an ignition device, was used as late as 1895 by the American, Hiram Percy Maxim, a graduate of Massachusetts Institute of Technology, when he built his first internal combustion engine for an automobile. Later, he would employ his considerable talent to designing and overseeing the production of electric cars for the Pope Manufacturing Company of Hartford, Connecticut, as outlined in *History of the Electric Automobile: Battery-Only Powered Cars.*[4]

Also contributing to the commercial success of the Otto engine was the discovery of oil in Pennsylvania in 1859. This event, aided and abetted by the American Civil War of 1861 to 1865, probably saved the whales from extinction. Providing the most nearly smokeless flame, whale oil, in the first half of the nineteenth century, had become preferred fuel for lamps and similar wick-fed devices. This flame, surrounded by a Fresnel lens, was used in lighthouses worldwide. By 1859, eight years after the first publication of Herman Melville's *Moby Dick*, whale oil had reached, in modern terms, a price of $55 a gallon,[28] an amount which caused concern to Professor Joseph Henry, then chairman of the U.S. Lighthouse Board.

Initially, of all the hydrocarbon fractions present in petroleum, the most valuable was kerosene. This fluid fueled the kerosene lantern, invented around the same time as the discovery of petroleum in Pennsylvania. Priced at only a few cents per gallon, kerosene rapidly replaced whale oil for lighthouses. Likewise, with the development of the kerosene lantern in 1860, this source largely replaced the candle for map reading by officers of the Union Army in the Civil War. On the other hand, the Confederates had a less well-developed infrastructure and con-tinued to use candles. General Robert E. Lee, after the third day of fighting at Gettysburg,

Pennsylvania, and the repulse by Union forces of Pickett's charge, for example, studied his map by candlelight in his tent to determine the best roads for the retreat from Gettysburg back to the Confederacy.

While kerosene was initially the sought-after fraction, gasoline was considered of little value. During the next few decades, oil company investigators learned more about the many hydrocarbons in petroleum. Oil was replacing tallow as a wheel and axle lubricant, naphtha was substituted for coal for steam-powered naval picket-boats shortly before the Spanish-American War of 1898, and petroleum tar was being used as a roofing material.

Originally using natural gas as a combustion agent, Otto soon learned of a more portable fuel known as gasoline. This application launched the Otto cycle engine, allied with the automobile, into worldwide use. Whereas the steam engine is uniquely attributed to the Englishman Thomas Newcomen by his patent of 1705, the invention of the internal combustion engine was finally decided in 1912 in a court decision, *Selden vs. Ford*, to be a social invention (i.e., invented by many).[4]

# References

1. "Hybrid Cars," *Electric Vehicles*, **65**, 2 June 1979, pp. 22–23.
2. "Hybrid Systems," *Electric Vehicles*, **67,** 1 March 1981, p. 15.
3. "Developing a Hybrid with Disc Motors," *Electric Vehicles*, **66**, 1 March 1980, pp. 18–21.
4. Wakefield, Ernest H., *History of the Electric Automobile: Battery-Only Powered Cars*, Society of Automotive Engineers, Warrendale, PA, 1994.
5. Shafer, William H., Commonwealth Edison Company, electric vehicle spokesman, 27 April 1982 (retired).
6. Hooper, F.C., personal communication, 17 August 1992. See also: "Clean Air Race Leads to New Electric Car," *Electrical World*, 15 February 1972, pp. 89–90.
7. Assbeck, F., Bidan, P., Marpinard, J.C., and Salut, G., "Automatic Control of a Hybrid Vehicle," Eighth International Electric Vehicle Symposium, Washington, DC, 1986.
8. "Hybrid Drive System," *Electric Vehicles*, **60**, 4 December 1974, p. 21.
9. Stephenson, R. Rhoades, *et al.*, "Should We Have a New Engine?" Technical Report, JPL SP 43-17, Vol. II, Jet Propulsion Laboratory, California Institute of Technology, Pasadena, CA, August 1975, Chapter 9, pp. 4-5.
10. Fersen, O.G.W., "Bosch Investigates Hybrids," *Automotive Industries*, April 1974. For more on German hybrid electric vehicles, see also: Kalberlah, A., "Electric and Hybrid Vehicles in Germany," SAE Paper No. 864898, Society of Automotive Engineers, Warrendale, PA, 1986.
11. Sariddakis, Nikolause, and Josefowitz, Willi, "Hybrid Petro-Electric Power System in the VW Golf Vehicle: Conception and Test Results," Eighth International Electric Vehicle Symposium, Washington, DC, 1986.
12. Wouk, Victor, "Two Decades of 'High Performance' EV Fleet Experience," Eighth International Electric Vehicle Symposium, Washington, DC, 1986.

13. Scott, David, and McCosh, Dan, "Automotive Newsfront," *Popular Science*, July 1992, p. 36. Also: "Opel Twin: Model Compact Car of Tomorrow," press release, Adam Opel, Germany, March 1992.
14. Scott, David, "Ford's Double Action Cruiser," *Popular Science*, January 1994, p. 51. See also: Bates, Bradford, "Getting a Ford HEV on the Road," *IEEE Spectrum*, July 1995, pp. 22–25.
15. "Chico, Das Neue Konzeptfahrzeug," Volkswagen press release, 1991.
16. "Latest Development Is Hybrid from Toyota," *Electric Vehicles*, **62**, 1 March 1976, p. 23.
17. Yamaguchi, Jack, "Toyota Prius," *Automotive Engineering*, January 1998, p. 30.
18. Nicholson, Leslie J., "New Technologies Will Be Able to Reduce Pollutants from Our Cars," *The Philadelphia Inquirer*, January 16, 1998. Also: "Toyota Gets Prius Orders 3.5 Times Initial Monthly Goal," *Kyodo News International*, 14 January 1998.
19. Lehotsky, Ronald M., "Powered by the Sun," personal communication, 31 July 1992. Also: Masashi Aihara, personal communication, 5 February, 1996.
20. Wood, Daniel B., "LA Launches Electric Car," *The Christian Science Monitor*, 11 December 1991, p. 12.
21. "LA 301 Technical Highlights," Clean Air International Designs Ltd., June 1992.
22. Letter from Count Moritz Jacobi to Michael Faraday, *Mechanics' Journal*, 1839, Vol. XXXII, p. 64.
23. "Hybrid Designed in Moscow," *Electric Vehicles*, **59**, 4 December 1973, p. 12.
24. Wouk, Victor, "Two Decades of 'High Performance' EV Fleet Experiences," Eighth International Electric Vehicle Symposium, Washington, DC, 1986.
25. *Encyclopedia Americana*, Vol. 15, p. 281.
26. *Scientific American*, December 1994, p. 55.
27. Asimov, Isaac, *Biographical Encyclopedia of Science and Technology*, Doubleday and Company, Garden City, NY, 1964.
28. Wakefield, Ernest H., *The Lighthouse That Wanted to Stay Lit*, Honors Press, Evanston, IL, 1992.

# CHAPTER 5

# Experiments with Non-Petro-Electric Vehicles

Since early times, humans have desired to travel using natural forces. As a result, the Persian-inspired magic carpet has arisen. The Greeks were particularly fertile in thought about such travel. Poliocretes carved *The Winged Victory of Samonthrace*. Mercury is a male with wings at his ankles. Icarius flew too near the sun, and his waxed wings melted. Pegasus was the winged steed of the Muses, who would carry the poet-rider to lofty inspiration. The Assyrians also had their Winged Deity. In the Christian era, winged angels hallow Jesus Christ. All can soar.

Seemingly to provide credence in modern times to these romanticisms have been the aircraft of aeronautical engineer, Dr. Paul B. MacCready. His creation, *Gossamer Condor*, was the first sustained heavier-than-air flight using only human energy. Following this *tour de force*, MacCready designed and built *Solar Challenger*, enabling a pilot to fly the 165 miles (265 km) from Paris to England powered solely from the sun. Solar-activated photovoltaic cells energized an electric motor which rotated a propeller.[1] More recently, MacCready was one of the key persons behind the General Motors Corporation solar-electric *Sunraycer*, a masterfully designed solar cell-battery powered motorcar.[2]

Somewhat allied in philosophy have been seven experiments with non-petro-electric vehicles: 1) Count Felix Carli's electric-spring powered tricycle of 1894, 2) the modern flywheel-electric vehicle (discussed in Chapter 6), 3) the General Electric battery-battery car, 4) the fuel cell-battery vehicle, 5) the inductively charged vehicle, 6) the turbine-battery, and 7) the recent solar cell-battery motorcar. Before assessing these experiments in this chapter, note that Table 5.1 lists a range of energy storage from electrostatic means, a capacitor, to nuclear fuel.[3] For the former means, Richard A. Karlin, writing on energy storage, notes that, on accelerating the charged capacitors only (not including the mass of the vehicle) to 30 mph (48 km/h), the capacitors contain only sufficient energy for approximately two accelerations.

In the case of nuclear fuels, the energy density is seen as approximately 100 million times the situation for electrostatic means. Presently, what is used in electric cars is, in many cases, the secondary battery in which the electrical energy is converted, on charging, into chemical energy (sulfuric acid from lead sulfate). On request, the sulfuric acid is reconverted to lead sulfate, yielding electricity. The efficiency of this process is approximately 75%. However, as stated repeatedly in the preceding chapters, lead-acid batteries are perceived to yield too little range and are too heavy for an ideal personal electric vehicle.

## TABLE 5.1
## VOLUME DENSITY OF ENERGY STORAGE SYSTEMS

|  | Horsepower-Hours per Ft³ | |
| --- | --- | --- |
|  | **Low** | **High** |
| Electrostatic | — | 0.006 |
| Magnetic | 0.001 | 0.08 |
| Gravitational | 0.008 | 0.2 |
| Mechanical | 0.01 | 0.8 |
| Phase Change | 0.01 | 100.0 |
| Primary Battery | 0.2 | 10.0 |
| Secondary Battery | 0.6 | 2.0 |
| Fuel Cell | — | 100.0 |
| Fuel | — | 400.0 |
| Nuclear | 60,000 | Very high |

Source: Ref. 3.

# Carli's Electric-Spring Tricycle

In the early period of competition for motorcars, the chief confutation against the electric vehicle was range limitation. To combat this nemesis, textile entrepreneur Count Felix Carli of Italy in 1894 was the first to offer a positive approach. His battery-spring tricycle, shown in

Figure 5.1, was equipped with "an impulsion box that is held in reserve...(containing) a system of rubber tension springs...(to) produce upon the axle an impetus...for a run of 160 ft (49 m)."[4] Carli's tricycle was almost surely the first hybrid electric vehicle and was of the parallel version. A second vehicle, which was a petro-electric type, would not appear until 1898 in America.

At the same time that Monsieur Pouchain was developing his first battery-only powered electric *phaeton* at Armentièrs in France, as described in Chapter 7 of *History of the Electric Automobile: Battery-Only Powered Cars*,[5] Carli was perfecting his multi-powered drive system. Historically, experimenters from Italy had been particularly conspicuous in observing new electrical phenomena. Alessandro Volta had assembled the first battery, heralding in a new scientific period. To cite another, Antonio Pacinotti had made two major observations: 1) that a ring-type assembly was desirable for an electric generator, and 2) this assembly could operate in two modes, either as a generator or as a motor. If mechanical power were applied to the assembly, electrical power would come forth. On the other hand, if electric power were introduced, mechanical power was made available. The Italians had a rich history to fulfill. Of Carli's carriage, a reporter writes[4]:

> This carriage was constructed at Castelnuovo in the power loom weaving establishment of Count (Felix) Carli, deputy to the Italian parliament. The Carli electric carriage is actuated by accumulators (batteries) of the Verdi type, this having been selected because it

*Figure 5.1 Count Carli's electric-spring tricycle. (Ref. 4)*

possesses a great specific capacity and can best resist the shocks always inevitable in a vehicle designed to run upon all sorts of roads. The battery consists of 10 elements, each having a capacity of 100 ampere-hours, say, 200 watt-hours. There is thus at one's disposal 2 kWh. The model employed weighs 11 lb (5 kg) and contains five plates. Under the conditions of normal discharge, the battery furnishes a current of 5 amperes, say, about a half-ampere per pound (0.5 kg). The rendering (voltage) alone drops from 97 to 63%, if we pass from a half to one ampere per pound of plates[a]...The vehicle weighs but 350 lb (159 kg) in running order.

The motor actuates the hind wheels directly by means of gearing. It absorbs about 550 watts, and the battery is capable of supplying it for four or five hours' trip (a range of 20 to 30 miles [30 to 50 km]). The excitation (magnetic field) is in derivation (parallel wound). The motor is capable of serving for the recharging of the accumulators, by the virtue of the well-known principle of reversibility. It is only necessary to apply a winch or a belt. There is a train of gearing between the axis of the motor and that of the wheels. By means of this gearing, it is possible, by turning a winch, to reduce the angular velocity of the motor from 1000 revolutions per minute to 100 or 30. On another side, a rheostat permits varying the angular velocity of the motor from 1000 to 300 revolutions per minute. It is thus possible to develop the greatest power corresponding to every speed, to run at slow speed upon ascending roads, at high speed upon declivities, etc.

For starting and for unforeseen obstacles on the way, recourse is had to an impulsion box that is held in reserve (multi-powered source). This consists of a system of rubber tension springs that are stretched by revolving a small wheel, even during the running of the carriage. When an energetic impulsion is necessary, the springs are relaxed by means of the foot and produce upon the axle an impetus equal to double the power of the motor itself and sufficient for a run of at least 160 ft (49 m).

The Carli establishment, under the able superintendency of Mr. F. Boggio, is constructing two types of this carriage, one of them simple and cheap and the other more elegant and more elaborate in detail. It is the second type that is represented in the figure...A few more improvements in accumulators and central stations will, in the charging of coach accumulators during the day and a part of the night, have an important market that will improve their annual rendering (earnings) as well as their present conditions of exploitation.

The Carli tricycle of 1894 had forward steering as opposed to Professors Ayrton and Perry's 1882 battery-only powered tricycle, which had rear wheel control.[4] Both groups used direct gearing to drive the wheels. In speed control, Carli, by changing gear ratios, utilized the principle of the transmission. He also had voltage speed control, as did the two professors. In short, Carli's electric carriage had features applicable to electric personal transportation in the

---

[a] Today, a battery with such internal resistance would be intolerable. See Chapter 11 in Ref. 5 to understand how this negative effect was mitigated.

protected and equitable environment of modern times. Regarding the last two paragraphs of the Carli report, essentially more than 100 years will have passed before the statements made by the reporter are seen as coming true, a time scale not realized at the time. Eventually, the waters seem to close over Carli. He is not heard again concerning vehicles, but his principle of making available both an inexpensive model vehicle and one more richly fitted is considered attractive marketing strategy by the modern automobile industry. Carli was first.

## Battery-Battery Automobiles

*The Wall Street Journal*[6] noted on 29 September 1993:

> The Clinton administration and the Big Three auto makers plan to unveil today a joint research program to triple the fuel efficiency of U.S. autos to an average of about 80 miles per gallon (about 30 km per liter) over the next decade...The vehicles will be comparable in size to today's cars but could use fuel cells or advanced energy storage systems rather than conventional engines.

This statement may be an early sign of a future demise of the more than a century of essentially an absolute reign of the internal combustion engine and its accompanying fuel, gasoline, in passenger motorcars.

One solution advanced to satisfy this statement is the battery-battery electric car in which two different kinds of batteries power the vehicle. The concept, it is believed, was first employed by the General Electric Company U.S.A. in its 1970 *Delta* electric vehicle shown in Figure 5.2 and in Chapter 21 of *History of the Electric Automobile: Battery-Only Powered Cars.*[5] General Electric used lead-acid and nickel-cadmium batteries, the latter providing power for acceleration and the former providing constant velocity. The battery-battery principle as described by R. Miles *et al.*[7] is as follows:

> A superior electric vehicle traction battery could be formed by a hybrid (multi-powered motorcar with) two batteries which take advantage of the good features of both battery technologies...a parallel combination is proposed which allows each battery to supply power depending on the type demand: high power for acceleration, and lower power for constant velocity.

Miles envisioned using an advanced lead-acid battery for high power in acceleration and a zinc-bromine battery for constant velocity.

*Figure 5.2. The General Electric battery-battery powered* Delta. *(General Electric Company)*

## Fuel Cell-Battery Powered Automobiles

Other approaches to increasing the range of electric automobiles are fuel cell technologies now under development and financed by the U.S. Department of Energy and others. A fuel cell is a galvanic cell that converts chemical energy to electric energy by combining a fuel and an oxidant (i.e., hydrogen and oxygen). A fuel cell, in the same way as the more common cells in a storage battery, produces direct current with a voltage under load of less than one volt. Hence, fuel cells must be connected in series to obtain higher voltages.

The conversion efficiency of chemical energy to electrical energy in a fuel cell is approximately 65 to 85%, a ratio higher than other more common, Carnot-type conversions, as illustrated in Figure 5.3. The advantages of fuel cells over electric cells commonly assembled in the conventional automobile battery, for example, are: 1) their electrodes last longer; 2) they operate for a longer time because the active material is continually replaced; 3) they have greater electrode surface per unit weight and volume; and 4) they can use readily accessible materials as fuel. These advantages, in part, are counterbalanced by the cost of the cells and auxiliary equipment, and for reasonably sized fuel cells, lower specific power.

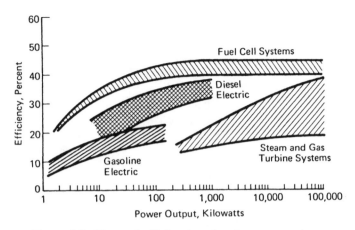

*Figure 5.3. Thermal efficiencies of various conversions.*
*(Ref. 8, Pratt & Whitney Division, United Technology Corporation)*

$$H_2 \text{ (a gas)} + 2OH^- \text{ (a liquid)} \rightarrow 2H_2O \text{ (a liquid)} + 2e^-$$

When oxidized at the anode, the fuel releases electrons that flow through the external load to the cathode where they reduce the oxidant. The reaction at the cathode is:

$$O_2 \text{ (a gas)} + H_2O \text{ (a liquid)} + 2e^- \rightarrow HO_2^- \text{ (a liquid)} + OH^- \text{ (a liquid)}$$

Figure 5.4 illustrates a Bacon or potassium hydroxide electrolyte type fuel cell utilizing hydrogen as a fuel source.[9]

Historically, Sir William R. Grove of England assembled the first fuel cell in 1839.[10] He would also suggest to Professor Moritz Jacobi, in service to Czar Nicholas I of Russia, the motive power (platinum-zinc electrode batteries) for the 1838 battery-powered electric boat, a 28-ft (9-m) craft which, transporting 12 people, traveled on the Neva River in St. Petersburg at 2.25 mph (3.6 km/h). The electric motor operated paddle wheels.[11] (Nicholas Roosevelt patented the side paddle-wheel concept in 1798 and made possible Robert Fulton's 1807 *Clermont*, the first commercially successful steamboat on American waters. The paddle-wheel patent was the genesis of the Roosevelt family fortune.) Surprisingly, interest in the fuel cell lagged until 1959 when Francis T. Bacon and J.C. Frost developed a 6-kW fuel cell.[12] Presently, a bus with fuel cells and batteries is traveling around the United States under the auspices of the U.S. Department of Energy.[13]

As reported by the U.S. Department of Energy in Ref. 13:

> ...the 30-ft (9-m) fuel cell transit bus is a heavy-duty, extremely low-emission electric vehicle that meets or exceeds all transit industry and Americans with Disabilities Act requirements. It is designed to operate in all climates, traffic conditions, and on the majority

99

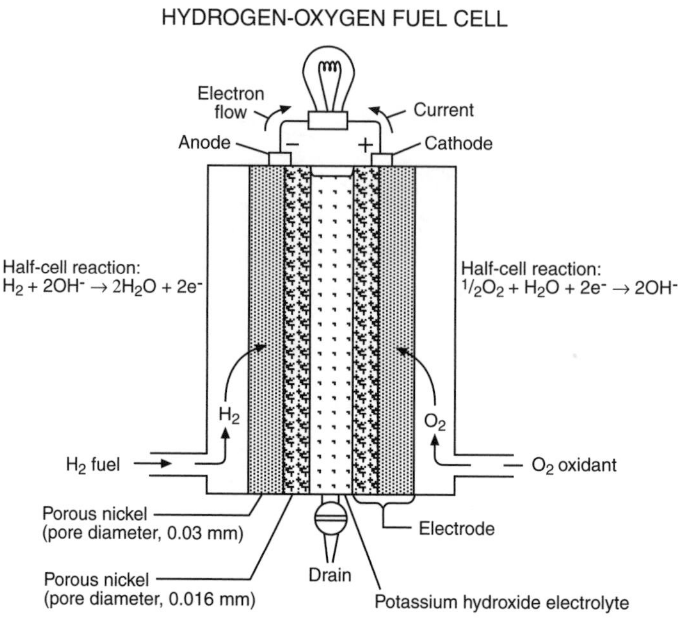

HYDROGEN-OXYGEN FUEL CELL

*Figure 5.4. An illustration of a Bacon or potassium hydroxide electrolyte type fuel cell utilizing hydrogen as a fuel source.  (Ref. 9)*

of transit routes.  Due to the fuel cell's high efficiency and low maintenance, the fuel cell bus is expected to have a lower life cycle cost than a low-emission diesel bus.  By combining a proven bus design with proven state-of-the-art fuel cell and electric drive technology, the fuel cell bus will provide a cleaner, quieter, and lower-cost form of mass transportation.

A fuel cell is a simple device that uses hydrogen and oxygen from the air, and combines them to form electricity and water vapor.  To meet the average transit bus's range and rapid refuel requirements, a liquid fuel, methanol, is reformed to provide the required hydrogen for the fuel cell.  The fuel cell bus employs the most highly developed type of fuel cell, a phosphoric acid fuel cell (PAFC).  This type of fuel cell is in commercial operation at sites around the world.  The 30-ft (9-m) bus uses a 50-kW PAFC system to provide 100% of the energy needed to complete a day's transit mission.  A battery pack is used to provide the peak power required for acceleration and hill climbing.  As an additional benefit, the battery stores the energy recovered from regenerative braking, resulting in reduced methanol consumption and extended battery life.

The drive system of the bus is seen to operate as a parallel system.

Figure 5.5 illustrates the fuel cell bus; Figure 5.6 is a block diagram of the fuel cell bus multi-powered system.  Table 5.2 compares the emissions from the fuel cell bus with diesel emissions standards for 1993.

*Figure 5.5.  The American fuel cell bus.  (U.S. Department of Energy)*

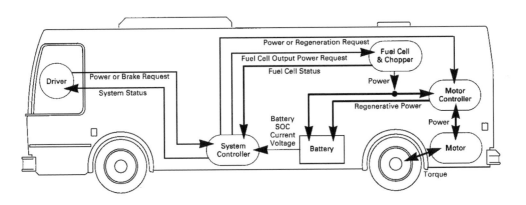

*Figure 5.6.  A diagram of the fuel cell bus multi-powered system.  (U.S. Department of Energy)*

101

TABLE 5.2
FUEL CELL VS. DIESEL EMISSIONS

|  | CO | $NO_x$ | HC | PM |
|---|---|---|---|---|
| 1993 Federal Emission Standard | 15.5 | 5.0 | 1.3 | 0.1 |
| 1998 Federal Emission Standard | 15.5 | 4.0 | 1.3 | 0.05 |
| Fuel Cell Bus | 0.35 | 0.01 | 0 | 0 |

All units in g/bhp-hr.
Source: U.S. Department of Energy.

# The Family of Fuel Cells

As reported by P.G. Patil *et al.* in Ref. 14:

> Alkaline fuel cell (AFC), phosphoric acid fuel cell (PAFC), proton exchange membrane (PEM) fuel cell, molten carbonate fuel cell (MCFC), and solid oxide fuel cell (SOFC). The characteristics and status of the several fuel cell technologies are summarized (in Tables 5.3 and 5.4):

TABLE 5.3
STATUS OF FUEL CELL TECHNOLOGIES

| Type of Fuel Cell | Operating Temp. (°C) | Fuel Compatibility | Coolant | Status |
|---|---|---|---|---|
| AFC | 25–100 | Pure $H_2$ | | 15-kW modules in use |
| PAFC | 200 | Dilute $H_2$ with <0.5% CO | Water, oil, or air | 100-kW tested |
| PEM | 80 | Dilute $H_2$ with <1 ppm CO | Water | 1-kW being tested |
| MCFC | 650 | $H_2$, CO | Air | 20-kW tested |
| SOFC | 1000 | $H_2$, CO, $CH_4$, $CH_3OH$ | Air | Tubular cells of 3-kW tested, monolith in laboratory stage |

Source: Ref. 14.

**TABLE 5.4**
**CHARACTERISTICS OF FUEL CELL SYSTEMS**

| Type of Fuel Cell | Current Density (mA/cm²) | | Power Density (kW/ft³) | | Start-UpTime (min) | | Dynamic Response |
|---|---|---|---|---|---|---|---|
| | Single Cell | Stack | Stack | System | Standby | Cold | |
| AFC | 5000 | 1000 | 10 | — | 1 | | Fast |
| PAFC | 500 | 240 | 3 | 0.5 | 15 | 300 | Slow |
| PEM | 4000 | 2000 | 20 | 0.8 | 1 | 5 | Slow |
| MCFC | 400 | 250 | 6 | 1.4 | 15 | 500 | Fast |
| SOFC | 2000 | — | 20 | 2.3 | 1 | 100 | Fast |

Source: Ref. 14.

Presently, the compact monolithic SOFC fuel cell also has interest. Argonne National Laboratory invented it in 1985. Continuing, P.G. Patil et al.[14] write:

> The monolithic fuel cell is a high-temperature device operating at 1000 degrees Centigrade. Its advantages over other fuel cells include high efficiency, high power density, and low cost materials, and it does not require an external reformer (to extract hydrogen from the liquid fuel) or battery pack for power peaking. A preliminary development schedule to build a 10-kW brassboard system followed by a 60-kW full-sized power source is shown (in Figure 5.7).

The potential improvement of both the PEM and solid oxide fuel cells compared with phosphoric acid fuel cells is shown in Figure 5.8. While the PAFC fuel cell is, at this writing, the most advanced system and indeed is operating in a city bus, the type that market forces will ultimately select for near-term passenger electric automobiles is now problematic. What about highway trucks also?

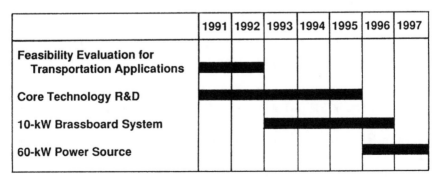

*Figure 5.7. Preliminary development schedule for monolithic SOFC vehicle power source. (Ref. 14)*

| Parameters | Proton-Exchange Membrane | Monolithic Solid Oxide |
|---|---|---|
| System Weight | 25% Less | 70% Less |
| System Volume | 40% Less | 80% Less |
| External Reformer | Still Required | Eliminated |
| Peaking Battery | Smaller | Eliminated |
| Cost | 25% Less | 75% Less |

*Figure 5.8. Potential improvement of advanced fuel cells compared with phosphoric acid fuel cells. (Ref. 14)*

Finally, for commercialization of fuel cells for electric cars, *The Wall Street Journal*[15] announced that the U.S. Department of Energy issued 30-month contracts of $13.8 and $15.0 million to Ford and Chrysler, respectively, to develop lightweight fuel cells with production prototype cars due *circa* 2004.

I first encountered fuel cells for motorcars in 1968 when I was invited to the New York headquarters of AT&T to meet with Union Carbide engineers (who were engaged in fuel cell work) and battery personnel from Bell Laboratories. The reason for the meeting was to discuss powering a fleet van for the Illinois Bell Telephone Company using a lead-acid battery providing variable-frequency and pulse-width-modulated, three-phase power to supply an induction motor, a concept I conceived independently in 1965. At that time, the methanol-oxygen fuel cells were perceived as uneconomical because of their estimated cost of approximately $2000/kW.

# Aluminum-Air Fuel Cell with Lead-Acid Battery Van

Another approach to increasing the range of the electric car is the aluminum-air fuel cell associated with conventional batteries. According to Dr. John H. Stannard of Alupower Canada Limited, the corporation has equipped a Chrysler Minivan with what the company calls an *Alupower Range Extender.* The vehicle was funded by Alcan International Limited and the Ontario (Canadian) Ministry of Energy. The article states[16]:

> Such a fuel cell, which has a mass of about 390 kg (860 lb), allows the range of an electric van to be extended from 75 km (47 miles) (on lead-acid batteries alone) to 300 km (186 miles) of mixed C cycle (0.54 km/cycle, or 0.34 miles/cycle) and 50 km/h (31 mph) driving. The range extender is based on technology for a refuelable aluminum fuel cell originally developed at Alcan's Kingston Research and Development Centre in Ontario, Canada. In 1988, this invention won Alcan a gold award in the prestigious Canada Awards for Business Excellence program.

The test vehicle for this program is a Chrysler Minivan purchased by Alupower. The van was converted to battery electric drive by Unique Mobility in Colorado. Onboard instrumentation allows instantaneous access to vehicle performance information such as motor, auxiliary, and controller chopper power consumption. Road and dynamometer tests were performed at Canada's Federal Vehicle Test and Emission Laboratory (VTEL) and Shell Canada Ltd.'s test track to determine vehicle energy and power requirements and driveline component efficiencies. The test results were used to determine power and energy requirements for the aluminum-air fuel cell.

Alupower has designed, built, and tested a 7.5-kW fuel cell in the Minivan. Together, the aluminum-air fuel cell and existing lead-acid batteries can power the vehicle for 300 km (186 miles). This range is calculated with 50% of the distance driven as C-cycles and 50% at a constant speed of 50 km/h (31 mph). This (program) offers a total driving time of nine hours. The fuel cell is installed above the electric drive motor, which is placed in the location formerly occupied by the internal combustion engine. An electrolyte heat exchanger is located at the mid-point of the vehicle under the floor, while the electrolyte tank is mounted behind the vehicle's rear axle. Passenger space is not affected in the conversion to range-extended operation. A 250-km (155-mile) demonstration run was done for the program in July 1991.

A transparent view of the Minivan is shown in Figure 5.9. Its characteristics are illustrated in Table 5.5. Figure 5.10 gives a volume comparison in stored energy of aluminum-air fuel cell and an equal amount of stored energy in lead-acid batteries. Refs. 17 and 18 provide more details about aluminum air fuel cells. In the final analysis, the market will sort out the energy storage system for electric cars.

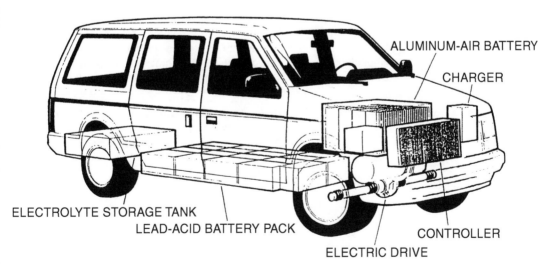

*Figure 5.9. A transparent view of the aluminum-air fuel cell-battery Minivan.*
*(Alcan International Limited)*

## TABLE 5.5
## CHARACTERISTICS OF THE ALUPOWER MINIVAN

| | |
|---|---|
| **Model Name:** | 6 kW aluminum-air prototype |
| **Manufacturer:** | Alupower Canada Limited, Kingston, Ontario, Canada |
| **Description:** | Chrysler Minivan converted to battery electric drive by Unique Mobility Inc., Englewood, Colorado, with 7.5 kW aluminum-air range extender, Alupower Canada Ltd. The range extender fuel cell is mechanically refuelable. |

| **Vehicle Dimensions:** | Length | 4.838 m | Curb Weight | 2535 kg |
|---|---|---|---|---|
| | Wheelbase | 3.025 m | GVW | 2875 kg |
| | Track | 1.578 m | Payload | 340 kg |

| **Electric Motor:** | Type | 108 V shunt wound d-c with separately excited field | | |
|---|---|---|---|---|
| | Power | 60 kW (80 hp) | Size | 0.28 × 0.5 m |
| | Weight | 140 kg | Transaxle | 5-speed manual |

| **Controller:** | Type | Solid-state, 0–100% control with field weakening |
|---|---|---|
| | Voltage | 108 V |
| | Current | 600 Amp |

| **Batteries:** | Type | Chloride 3ET-205 lead acid |
|---|---|---|
| | Number | 18 × 6 volt modules |

| **Onboard Charger:** | Output Current | 16 amp/20 amp/40 amp at 115 VAC |
|---|---|---|
| | Charge Time | 12 hrs/6 hrs |

| **Range Extender:** | Type | 6-kW nominal aluminum-air fuel cell, 56 cells in series (7.5 kW gross power output) |
|---|---|---|
| | Output | Approximately 100 amperes at 60 volts charging lead-acid batteries via 95%-efficient dc-dc converter |
| | Subsystems | Cell stack, heat exchanger, electrolyte recirculation and remote electrolyte reservoir |
| | Capacity | Approximately 70 kWh at 9-hr discharge rate |
| | Dimensions | Mass is 390 kg; volume is 320 l; as-built prototype installed in Minivan |

| **Performance:** | Typical Range | 75 km (lead-acid batteries only; 50% C-cycle, and 50% constant 50 km/h) |
|---|---|---|
| | | 250 km (as aluminum/air hybrid—demonstrated) |
| | Maximum Speed | 110 km/h—demonstrated |

Source: Alcan International Limited.

*Figure 5.10. Energy volume comparison of aluminum-air fuel cell and lead-acid battery. (Alcan International Limited)*

Another means of extending the range of electric vehicles, in this case on repetitive routes, is to employ inductive charging of the battery.

## Noncontact Inductively Charged Moving Vehicles

A discovery of the indomitable and early nineteenth century scientist Michael Faraday was employed by personnel of the University of California since the mid-1970s to charge the battery of an electric vehicle.[b] Reported by R.E. Parsons *et al.* of the Lawrence Berkeley Laboratory and from the Lawrence Livermore National Laboratory, mutual induction was employed for charging.[19] Using the above development, the Santa Barbara Bus Project was initiated.[20] The purpose was to charge the battery of a bus while the vehicle was en route to its destination. To this end, a long primary winding of a transformer was placed in the roadway. This element was energized from the electric utility, but energized by a relatively high frequency to assist in power transfer across the air gap to the secondary of a transformer, borne by the bus. Electric current in the secondary was rectified and fed to either the battery or the motor of the bus. The schematic of the Santa Barbara bus project is shown in Figure 5.11.[20]

---

[b] The early work of Professor Francis Schwartz of the University of Leyden was cited by Dr. Victor Wouk, April 1994.

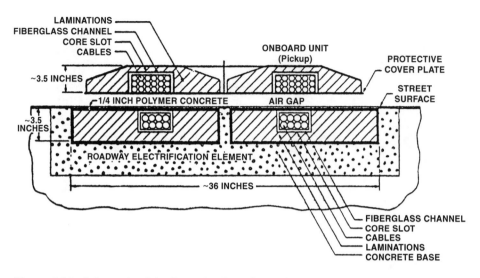

*Figure 5.11. Schematic of the Santa Barbara bus inductive energy transfer. (Ref. 20)*

Later, in Canada, P. Choudhury *et al.* of Queen's University applied these developments to the electric automobile.[21] Figure 5.12 exemplifies the Canadian approach; Table 5.6 gives their specifications and design guidelines. Note the 50-kW power transfer, at an efficiency of more than 80%. A unique contribution of the Canadians was to employ a "magnetic brush" mounted on the underside of the vehicle-borne secondary to reduce the effect of the air gap, as reported by D.L. Atherton, C. Welbourn, and M.G. Daly[22,23]:

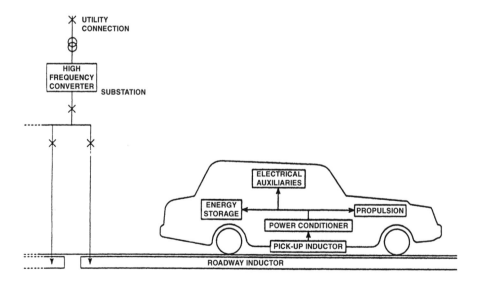

*Figure 5.12. Schematic of the energy transfer and power conditioning system. (Ref. 21)*

Flexible strips of ferromagnetic material would occupy the greater portion of the air gap and could even 'brush' along the primary inductor, thereby substantially reducing the reluctance and required magnetizing current of the E-core transformer magnetic circuit.

It goes without saying that charging by mutual induction could also be used for a single stationary vehicle.

### TABLE 5.6
### SPECIFICATION AND DESIGN GUIDELINES

| | | |
|---|---|---|
| **Vehicle Pick-Up Indicator:** | Power | 50 kW (maximum) |
| | Weight | <200 kg |
| | Length | <2 meters |
| | Winding Current Density | <4.5 A/mm² |
| **Highway Delivery Inductor:** | Block Length (Approx.) | 100 meters |
| | Winding Current Density | <2.5 A/ mm² |
| **Overall Requirements:** | Airgap | 5 cm |
| | Efficiency | >80% |
| | | for min. highway operation |
| | | (3 vehicles/highway block) |

William H. Shafer, long-term employee of Commonwealth Edison Company of Chicago and now retired, refurbished the above concept which might be considered a variation of a hybrid electric vehicle. The kernel of his idea is to use an underground power source to inductively charge the battery through an air-gap and a suitable loop to the battery while the passenger bus loads riders.[c,21] Shafer's approach, based on the 1984 experiments reported in the Society of Automotive Engineers (SAE) Paper No. 830350, is for the bus to carry a lesser battery load but charge the batteries by electro-magnetic linkage through mutual inductance coils. Buried in the pavement at each passenger pickup point would be an inductive coil energized from the local utility network but only with a presence of an autobus. When the bus was centered above the coil in the pavement, the inductive coil of the bus would be automatically lowered to reduce the air-gap between the two coils. Then an automatic transfer of a-c electric energy (400 Hertz has been used) would occur through the intervening air-space to an integrated circuit associated with the bus battery. In the short time necessary for passenger boarding, a considerable amount of electric energy would be absorbed by the bus battery through a suitable circuit. Thus, at every city block or two, the battery charge would be replenished. With full battery charge alone, the bus would have sufficient energy to travel approximately 5.6 miles (10 km). Therefore, the vehicle would have some independence from the charging network.

---

[c] Such a charging plan, plus solar cells, was also postulated in 1987 for the *Heliobat* for serving personnel at the proposed Chicago World's Fair of 1992, as cited in Chapter 11.

Placing numbers in this system, Shafer reports earlier Chicago 50-passenger trolley buses required 6.2 kWh/mile for operation. This energy requirement was divided: 3.1 for propulsion, 3.2 for resistance heating, and 0.2 for lighting and auxiliaries.[24] If, for the bus, the passenger boarding street charging system could transfer power to the bus at a rate of 46 kW, as reported for the cited similar Santa Barbara, California, prototype system, then for every minute of standing time the bus battery would be charged with approximately 0.8 kWh. If there are eight city blocks per mile, passengers were boarded at each block, and passenger boarding time were one minute, then the battery would be charged more than 6 kWh/mile, an amount ample for the 50-passenger bus to rely on only passenger-stop charging.

# Summary

I have cited here several multi-powered systems for electric automobiles. In addition to the petroleum-fueled cars, the dual-powered cars must compete with forthcoming vehicles bearing batteries of higher specific energy and also the interchange battery concept. This latter approach, pioneered by Jeantaud and Brault in the 1895 Paris-Bordeaux-Paris race and described in Chapter 15 of *History of the Electric Automobile: Battery-Only Powered Cars*,[5] was modernized by Robert McKee *et al.* as discussed in Chapter 21 of Ref. 5. If the highly developed infrastructure (service station) of the internal combustion car would offer battery interchange, the niche for the hybrid electric car appears small. On the other hand, if the electric utility industry, with almost flat growth projected in the 1990s, wishes to provide electric energy to transportation, it might indeed employ petro-electric cars in its own fleets. With the present dominance of the petroleum-fueled automobile, it appears that the electric car can flourish only in special niches unless a government regulation specifies otherwise. However, these niches are widening with the growing concern about air pollution. Barring unforeseen circumstances, it can be argued that this trend will be true. Air pollution in cities is the new force, which is encouraging present motorcar manufacturers worldwide to place more emphasis on the nearly pollution-free electric automobile and its cousin, the hybrid electric vehicle.[25] Among the hybrid electric cars also bearing close scrutiny are the flywheel-electric cars, the theory and present status of which are discussed in the next chapter.

# Note 1—The General Motors Fuel Cell-Battery Powered Van

In 1967, a fuel cell-battery powered van was built by General Motors under the general direction of P.D. Agarwal, as shown in Figure 5.13. Note the liquid hydrogen and oxygen fuel tanks. The a-c drive system for this vehicle is described in Chapter 19 of *History of the Electric Automobile: Battery-Only Powered Cars*.[5]

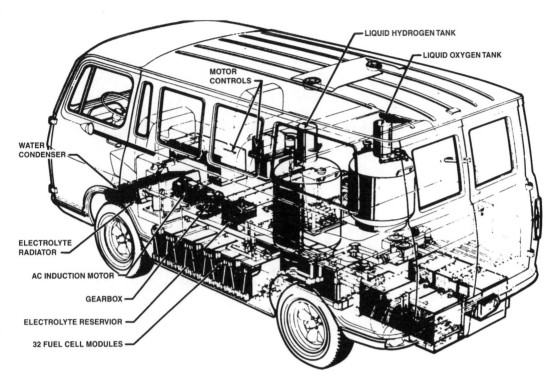

*Figure 5.13. GM fuel cell-battery powered van with three-phase, a-c drive using variable-frequency, pulse-width-modulated controls. (General Motors)*

# Note 2—The Canadian Hydrogen-Fueled Bus

*AutoWeek* of 8 May 1995[26] states that General Motors issued a $4.3 million contract to Ballard Power Systems of North Vancouver, British Columbia, Canada, to produce fuel cells for an automotive engine for "testing and integration by GM into a hybrid fuel cell engine. This engine will be (placed) into a prototype for testing." In addition, with financial support from the British Columbian and Canadian federal governments, Ballard has delivered a 21-passenger, hydrogen-fueled, proton exchange membrane (PEM), fuel cell powered bus, shown in Figure 5.14 with the characteristics shown in Table 5.7.[27]

According to Bonne W. Posma, President of Saminco Inc., his company manufactured the drive system for not only on the Ballard bus but also the U.S. Department of Energy methane-fueled bus mentioned previously.[28] In addition, Posma writes[27]:

### Why Batteries Won't Do

A large, family-sized car traveling at 60 mph (97 km/h) consumes energy at the rate of 40 kW (including power for accessories such as heater, air conditioner, power steering and brakes, head lamps, and oversized stereo). At this speed, it would take five hours to cover

*Figure 5.14. The Canadian hydrogen-fueled bus. (Saminco Inc.)*

its range of 300 miles (483 km), and thus, the gas tank contains 200 kWh of usable energy (5 hours × 40 kW). Assuming consumption at 20 miles/gallon, the tank would hold 15 gallons.

When filling up, gasoline is pumped at the rate of 10 gallons/minute, so it takes 1.5 minutes to completely fill the tank (0.025 hours). During pumping, energy is transferred at the rate of 8 mW (200 kWh/0.025 hours). (mW stands for megawatt.) It would only take 75 cars filling up simultaneously to transfer a total amount of energy at the rate of 600 mW (8 mW × 75), which is equivalent to the entire output of a medium-sized power station.

If we provide a perfect secondary (rechargeable) battery for a vehicle, at 200 V, whose charge/discharge efficiency (were) 100%, and one that could be fully charged in only 1.5 minutes, to make it equivalent in convenience to filling up a gas tank, then we would have to supply the enormous current of 40,000 A for 1.5 minutes for a full recharge. As was pointed out before, the entire output of the 600 mW power plant would have to be diverted to satisfy the power demands of as few as 75 cars. Clearly, this would place untenable demands on the area's electric...system.

## TABLE 5.7
## CHARACTERISTICS OF THE PEM FUEL CELL POWERED TRANSIT BUS

| Full Load Performance | |
|---|---|
| Gradability | Start on 20% Grade |
| | Maintain 30 km/h (20 mph) on 8% Grade |
| Acceleration | 0 - 50 km/h (0 - 30 mph) in 20 sec |
| Top Speed | 70 km/h (45 mph) |
| Range | 150 km (94 mi) |
| Meets Urban Mass Transit | |
| Authority (UMTA) Performance Criteria | |

| Fuel Cell Power Plant | |
|---|---|
| Fuel Cell Type | 24 Ballard PEM MK5D stacks |
| Arrangement | 3 parallel strings of 8 in series |
| Voltage Range | 160 - 280 Vdc |
| Gross Power | 120 kW |
| Operating Fuel Pressure | 207 kPa (30 psig) |
| Operating Air Pressure | 207 kPa (30 psig) |
| Operating Water Temperature | 70 - 80°C (160 - 175°F) |
| Emissions | Zero |

| Fuel System | |
|---|---|
| Fuel Type | Gaseous hydrogen |
| Storage Pressure | 20680 kPa (3000 psig) |
| Delivery Pressure | Regulated to 207 kPa (30 psig) |
| Recirculation Method | Ejector |
| Storage Method | Aluminum / Fibreglass cylinders |
| Storage Capacity | 125 scm (4410 scf) |

| Air Compression System | |
|---|---|
| Compression | Up to 207 kPa (30 psig) |
| Delivery Capacity | 7362 lpm (260 cfm) |
| Drive Motor | Brushless DC motor and controller |
| | 24 kW @ 5000 rpm, 140 - 200 Vdc |

| Traction Drive System | |
|---|---|
| Motor Type | DC shunt field motor |
| Controller Type | 2 quadrant DC/DC IGBT chopper |
| Controller Frequency | 400 Hz |
| Controller Input Voltage | 160 - 280 Vdc |
| Motor / Controller Efficiency | 80% |
| Power Output (Continuous) | 80 kW (107 HP) |
| Motor: | |
| Base Speed | 1900 rpm |
| Field Weakened Speed | up to 3000 rpm |

| Electrical System | |
|---|---|
| Main DC Link Voltage | 160 - 280 Vdc |
| Starting Battery and Charger: | |
| Voltage | 144 Vdc |
| Type | 12 Lead / Acid |
| Capacity | 20 Ah |
| Auxiliary Battery and Charger: | |
| Voltage | 12 Vdc |
| Type | Lead / Acid |

| Cooling System | |
|---|---|
| Fuel Cell Power Plant | Water cooled |
| Radiator | 1.39 m$^2$ (15 ft$^2$) |

| Control System | |
|---|---|
| Hardware | Intel 286AT computer controller |
| Software | Custom designed |

| Safety Features | |
|---|---|
| Monitoring | Low voltage shutdown |
| Gas Detection | Hydrogen detectors |
| Electrical | Single / double fault |

| Environmental | |
|---|---|
| Operating Temperature | 3 - 40°C (35 - 105°F) |
| Storage Temperature | 3 - 55°C (35 - 125°F) |
| Vibration | As per MIL-STD-180E |
| | Category 1 |

## Note 3—*Mirai 1*

Another experimental fuel cell-battery vehicle is Sanyo's *Mirai 1* ("future").[29] Dennis Normile writes after speaking with Sanyo Manager Akio Takeoka:

> *Mirai 1*...shows off *three* energy-related technologies...solar cells and nickel-cadmium batteries in addition to the fuel cell (phosphoric-acid type). The solar cells and/or the fuel cell charge the batteries, which drive a brushless direct-current motor.
>
> The car's top speed is just 62 mph (100 km/h), and it can run only for about two hours on a full charge. But continued charging by the solar panels and/or fuel cell extends that. The fuel cell generates electricity through a chemical reaction between hydrogen and air, a form of reverse hydrolysis; the only emissions are water vapor and air. Takeoka believes a fuel cell powered car would meet California's no-emission requirements for autos, which are set to take effect in 1998.
>
> Fuel cells are not likely to displace batteries completely. 'It takes a fuel cell several minutes to come up to full power, and then it's better to run it at constant output,' according to Takeoka. Sanyo envisions fuel cells being paired with batteries. The latter would provide the variable power needed for typical driving.
>
> The potential advantage of a fuel cell could come from vastly reducing the number of batteries required and being able to recharge on the go. But to make this worthwhile, Sanyo faces the same problem as the battery makers—getting more power out of a smaller, lighter, less-expensive fuel cell package. At present, the energy density of the fuel cell system, about 3.8 watt-(hours) per pound (8.3 watt-hours/kg) is less than one-third that of lead-acid batteries. Although it has a long cycle life, the fuel cell costs far more.

Again, the Achilles' heel of the electric car reveals itself.

## Note 4—The Chrysler Fuel Cell Powered Vehicle

Another approach to the hybrid electric automobile is described below.[30] "A team headed by Pentastar Electronics (a Chrysler Company) completed the first 15 months of a planned 30-month contract ($15 million, 20% cost-shared) to develop a fuel cell propulsion system using a design-to-cost approach. Subcontractors include Chrysler Liberty of Madison Heights, Michigan, and an AlliedSignal team led by AlliedSignal Aerospace Systems and Equipment of Torrance, California, that also includes AlliedSignal Automotive of Southfield, Michigan, and AlliedSignal Research and Technology of Morristown, New Jersey.

Efforts during fiscal year 1995 focused on the completion of a vehicle conceptual design and development of the fuel cell power system. In a report to be published by the Department, a conceptual design is presented based on the initial selection of the Chrysler sedan. (This vehicle would be equipped) with a 50-kW (gross) proton exchange membrane (PEM) fuel cell stack as the primary power source, a battery-powered load leveling unit for surge power requirements (100 kW total for acceleration), an onboard high-pressure gaseous-hydrogen storage system, a system to manage the hydrogen and air supplies for the fuel cell stack, and electronic controllers to control the electrical system (see Figure 5.15). As part of the conceptual design effort, investigation of onboard hydrogen storage technologies and assessments of hydrogen fueling infrastructure and safety requirements were also completed, with reports issued by Pentastar."

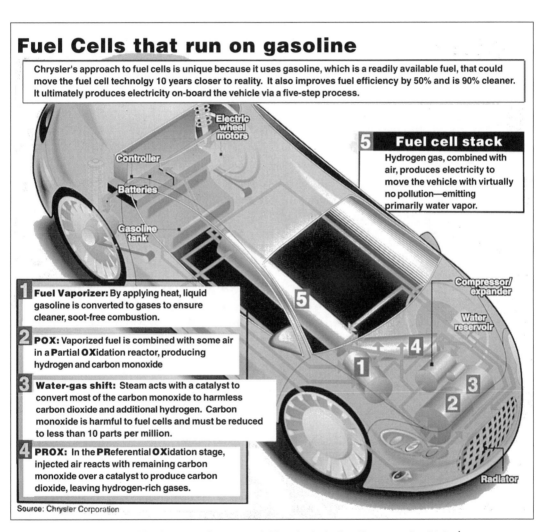

# Fuel Cells that run on gasoline

Chrysler's approach to fuel cells is unique because it uses gasoline, which is a readily available fuel, that could move the fuel cell technolgy 10 years closer to reality. It also improves fuel efficiency by 50% and is 90% cleaner. It ultimately produces electricity on-board the vehicle via a five-step process.

Electric wheel motors

Controller

Batteries

Gasoline tank

**5 Fuel cell stack**
Hydrogen gas, combined with air, produces electricity to move the vehicle with virtually no pollution—emitting primarily water vapor.

Compressor/expander

Water reservoir

**1 Fuel Vaporizer:** By applying heat, liquid gasoline is converted to gases to ensure cleaner, soot-free combustion.

**2 POX:** Vaporized fuel is combined with some air in a **P**artial **OX**idation reactor, producing hydrogen and carbon monoxide

**3 Water-gas shift:** Steam acts with a catalyst to convert most of the carbon monoxide to harmless carbon dioxide and additional hydrogen. Carbon monoxide is harmful to fuel cells and must be reduced to less than 10 parts per million.

**4 PROX:** In the **PR**eferential **OX**idation stage, injected air reacts with remaining carbon monoxide over a catalyst to produce carbon dioxide, leaving hydrogen-rich gases.

Radiator

Source: Chrysler Corporation

*Figure 5.15. Concept drawing of Chrysler's fuel cell powered vehicle.[d]*
*(A.C. Lieber, Scott Fosgard, Chrysler Corporation)*

[d] See also: Wilson, Kevin A., "The Fuel Cell Sell," *AutoWeek,* 3 March 1997, pp. 12–14.

# References

1. *GM Sunraycer Case History,* Society of Automotive Engineers, Warrendale, PA, p. 8.
2. Wilson, Howard G., MacCready, Paul B., and Kyle, Chester R., "Lessons of Sunraycer," *Scientific American,* March 1989, pp. 90–97.
3. *Battelle Technical Review,* **19**, 4 April 1966, p. 10.
4. *Scientific American,* **LXX**, 16, 21 April 1894, p. 251.
5. Wakefield, Ernest H., *History of the Electric Automobile: Battery-Only Powered Cars*, Society of Automotive Engineers, Warrendale, PA, 1994.
6. Davis, Bob, "White House and Auto Makers Prepare Joint Effort to Triple Fuel Efficiency," *The Wall Street Journal,* 29 September 1993, p. A9.
7. Miles, R., *et al.,* "All Electric Hybrid Power Source for Electric Vehicles," Tenth International Electric Vehicle Symposium, Hong Kong, 3–5 December 1990, pp. 777–783. See also: Laumeister, B.R., "Experimental Electric Vehicle Being Developed by General Electric," SAE Paper No. 680430, Mid-Year Meeting, Detroit, MI, 20–24 May 1968, Society of Automotive Engineers, Warrendale, PA, 1968.
8. Appleby, A.J. and Foulkes, F.R., *The Fuel Cell Handbook*, Van Nostrand Reinhold Co., 1968, p. 1117.
9. *Encyclopedia Americana,* Volume 12, Grolier, Danbury, CT, 1988, p. 146.
10. Grove, W.R., "On the Voltaic Series and the Combination of Gases by Platinum," *Phil. Mag.,* **111,** 14, 1839, p. 127.
11. *Scientific American,* **XLVII**, 16 December 1882, p. 388.
12. Mantell, Charles Letnam, *Batteries and Energy Systems,* 2nd ed., McGraw-Hill, New York, 1983.
13. *Fuel Cell Bus,* U.S. Department of Energy, Office of Transportation Technology, 1000 Independence Ave. S.W., Washington, DC 20585, April 1994.
14. Patil, P.G., Kost, R.A., and Miller, J.F., "U.S. Research and Development Program on Fuel Cells for Transportation Applications," Tenth Electric Vehicle Symposium, Hong Kong, 1990, pp. 657–669.
15. *The Wall Street Journal,* 13 July 1994, p. B2.
16. Fitzpatrick, N.P., personal communication, 3 May 1993.
17. Parish, D.W., *et al.,* "Demonstration of Aluminum-Air Fuel Cells in a Road Vehicle," SAE Paper No. 891690, Conference and Exposition on Future Transportation Technology, 7–10 August 1989, Vancouver, British Columbia, Canada, Society of Automotive Engineers, Warrendale, PA, 1989.
18. Rudd, E.J., "The Development of Aluminum Batteries for Electric Vehicles," SAE Paper No. 891660, Conference and Exposition on Future Transportation Technology, 7–10 August 1989, Vancouver, British Columbia, Canada, Society of Automotive Engineers, Warrendale, PA, 1989.
19. Parsons, R.E., "Program on Advanced Technology for the Highway (Project Report No. 3)," Phase I—Final Report, December 1987.
20. "Santa Barbara Electric Bus Project, Prototype Development and Testing Program, Phase 3B—Final Report," System Control Technology, Inc., September 1984.

21.  Choudhury, P., *et al.,* "Inductive Power Transfer to Highway Vehicles," SAE Paper No. 891706, Conference and Exposition on Future Transportation Technology, 7–10 August, 1989, Vancouver, British Columbia, Canada, Society of Automotive Engineers, Warrendale, PA, 1989.

22.  Atherton, D.L., and Welbourn, C., "A Rotating Drum Test Rig for the Development of Pipeline Monitoring Tools," *CSNDT Journal,* September 1985, pp. 50–56.

23.  Atherton, D.L., and Daly, M.G., *Finite Element Calculation of Magnetic Flux Leakage Detector Signals,* NDT International, Vol. 20, No. 4, 1987, pp. 235–238.

24.  Shafer, William H., personnel communication, June 1988.

25.  "Young Stylists Tackle the Hybrid Car," *Electric Vehicles,* **67**, 2 June 1981, p. 4. See also: Zetsche, Dieter, "The Automobile: Clean and Customized," *Scientific American*, September 1995, pp. 102–106.

26.  *AutoWeek*, 8 May 1995, p. 9.

27.  Posma, Bonne W., *Fuel Cell Powered Vehicles*, Saminco Inc., Fort Myers, FL, 11 January 1993.

28.  Posma, Bonne W., personal communication, 26 April 1995.

29.  Normile, Dennis, "Fuel Cell on Wheels," McCosh, Dan, ed., *Popular Science*, August 1992, p. 29.

30.  19th Annual Report to the U.S. Congress for Fiscal Year 1995.

# CHAPTER 6

# The Flywheel-Electric Vehicle

Flywheels were probably first used as an energy storage element in the potter's wheel, an instrument with a history of possibly 5000 years. The spindle or shaft of the potter's wheel is vertical. Bearings are suitably located. The moist clay to be shaped is placed on a head rigidly linked by the spindle to a flywheel near floor level. Energy is transferred to the flywheel by a kick of the operator's foot which forces the head bearing clay to rotate, thereby allowing the potter to shape the clay with his hands. The energy pulse imparted by the kick also was recognized by German experimenter Nicholas Otto in developing the four-cycle internal combustion engine in the mid-1870s. (See Note 2 in Chapter 4 for more about the Otto engine.) Like the potter, Otto grasped the desirability of a flywheel to smooth power flow from the energy pulses resulting from the explosions within the cylinder of the internal combustion engine. All gasoline-powered cars today bear a flywheel combined with the electric starter, and most modern punch presses possess a flywheel storing inertial energy. Therefore, the flywheel is seen as a ubiquitous means of energy storage.

The first use of a flywheel as a means of energy storage for extending the range of a vehicle probably was in a Swiss bus.[1] At passenger loading stations, an electric motor takes power from an overhead source and re-energizes the flywheel. Figures 6.1 and 6.2 illustrate this principle.[2] If such a drive system is indeed to be a hybrid electric vehicle, there must be another energy storage system.

While energy posited in a flywheel had seen recent technical papers,[3] one of the first presentations to reach a wide audience was by Richard F. Post and Stephen P. Post, physicists at the University of California, Berkeley.[4] The article advanced the employment of mechanical energy storage in a series of locally distributed flywheels, each interconnected with an electric motor/generator. This system would serve as an electric load-leveling device, mitigating an

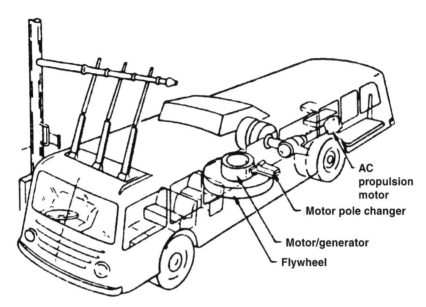

*Figure 6.1. Flywheel-powered passenger bus. (U. S. Energy and Development Administration)*

AC propulsion motor

Motor pole changer

Motor/generator

Flywheel

*Figure 6.2. English tram.[2] "Could electric transport be the answer to city center pollution?...Wessex Branch members will hear (of) one alternative—a no-wire mini-tram which is electrically powered but stores energy in a flywheel, removing the need for overhead wires. A new, full-scale production vehicle, above, is presently in service at Parry People Movers' demonstration track in the West Midlands." For additional information, see Note 3 at the end of this chapter.*

electric utility problem, load-leveling, which is discussed in Chapter 24 of *History of the Electric Automobile: Battery-Only Powered Cars*.[5] In day/night load-leveling, the flywheel, through a motor, would absorb excess electrical energy from the central generating station at night. During the day, which is a period of high electricity usage, the flywheel would yield inertial energy to the associated generator, thus providing electric power to the electrical distribution system and supplementing the output of the central station. This plan holds reminiscences of the more common water electrical pump-storage system. See Note 1 at the end of this chapter for mention of the Commonwealth Edison Company and Argonne National Laboratory flywheel system.

Although such application of the flywheel was the main thrust of the Posts' article, the use of the flywheel in electric cars also was explored. The Posts related that the flywheel, unlike the lead-acid battery, could be rapidly supplied with energy.[a] Moreover, in regenerative braking of an automobile, the transfer of energy of motion from the car to the flywheel and back to the car on acceleration had an efficiency of approximately 95% compared to a similar transfer in and out of a battery of 50 to 75%. What additionally triggered interest in the flywheel as a means for energy storage was the recent availability of fibers with high tensile strength capable of withstanding centrifugal forces associated with high rotational speeds. The quantity of energy stored in a flywheel is dependent on the mass density and the speed of rotation of the wheel: the former to the first power, and the latter to the second power. Hence, high energy storage is proportional to the rotational speed squared. To minimize air resistance, the need for vacuum sealing of the flywheel and the associated motor-generator was emphasized. Magnetic bearings also are required.

## Flywheel—Institut fur Kraftfahrwesen

A company in Germany, the Institut fur Kraftfahrwesen (IKA), has chosen a flywheel assist in a hybrid electric vehicle. As R. Rhoads Stephenson *et al.* write[6]:

> The IKA hybrid has a three-component storage system consisting of liquid fuel, a battery, and a flywheel. The flywheel is geared directly to the output shaft of a 20-hp single rotor Wankel engine. The engine shaft is connected to one side of a differential gear assembly, and an 11-hp electric traction motor drives the other side.[7] The combined output is taken from the differential carrier and fed through a transmission to the drive wheels. A block diagram of the configuration is shown in Figure 6.3. The flywheel weighs approximately 100 lb (45.36 kg) and spins at a maximum speed of 18,000 rpm. At this speed, the total stored energy is 468,000 ft-lb (175 watt-hours).

---

[a] The Posts failed to mention demand charges, nor did Aronson in *History of the Electric Automobile: Battery-Only Powered Cars*[5] emphasize this inherent negative and expensive factor in all rapid-charge schemes. A demand charge, over and above the quantity of energy used, is levied by a utility to help defray the capital cost for providing electricity to an outlet with irregular use.

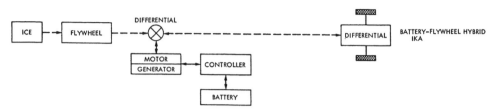

*Figure 6.3. The IKA drive system. (Ref. 6. R. Rhoads Stephenson, Jet Propulsion Laboratory)*

Discharging the flywheel to half this energy level (87 watt-hours) is equivalent to the kinetic energy of the 4630-lb (2100 kg) Volkswagen van in which the system has been tested at a speed of 28 mph (45 km/h).[b]  R. Rhoads Stephenson *et al.* continue[6]:

> The flywheel-differential-motor combination acts as an electric torque converter, permitting the output torque converter to be regulated from zero to an amount greater than that produced by the engine.  This (event) occurs even though the engine is physically connected in series with the other power-train components.  The characteristics of the system differ from those of the conventional hydrodynamic torque converter in that power conversion takes place at all engine and output shaft speeds and torques so long as the average engine power exceeds the average load power.  The flywheel absorbs load fluctuations, thereby permitting the engine to operate in a low response mode.  Rapid power response is provided by the electric motor, which reacts against the flywheel inertia and extracts or augments its energy, as well as that of the battery, during transient load changes.  To compensate for variations in average road conditions, engine speed may be adjusted over a relatively narrow range to maintain the battery charge state within an acceptable band.
>
> The system has been tested under simulated urban driving conditions.  Test results indicated that a 45% reduction in fuel consumption can be achieved for a three-stop-per-mile profile in which cruise speeds reach approximately 30 mph (48 km/h).  Fourteen percent of the fuel consumption reduction is attributed to regenerative braking, and the remaining 31% is provided by improved engine loading.  The overall fuel economy improved from 9.4 to 11.7 mpg for the 4 to 5 km/l conventionally powered van to 18.2 to 22.4 mpg for the hybrid conversions.  During acceleration tests, a speed of 30 mph (48 km/h) was reached in 10.8 seconds.  The system has been designed as a power plant for a much smaller vehicle and, although not reported, it is expected that the large drag coefficient, frontal area, and weight, in combination with the small engine of the experimental van installation, (will) severely limit sustained high-speed performance.

---

[b] Or cruise at 20 mph (32 km/h) for 0.5 mile (0.8 km)—see Figure 3.5 in Chapter 3.  If the 100-lb (45-kg) flywheel is spun at 100,000 rather than 18,000 rpm, using the figures above, the stored energy is:  $(100,000 \div 18,000)^2 \times$ 175 watt-hours = 5.3 kWh, an amount capable of driving the General Motors *Impact* approximately 46 miles (74 km) at 55 mph (89 km/h).

## Flywheel Work at Other Sites

The energy storage capability of flywheels has been investigated at other sites. Figure 6.4 illustrates several types of flywheels considered for vehicle use, and a high-speed flywheel energy storage system developed by F.J.M. Thoolen of the Netherlands is shown in Figure 6.5.[8] Note the use of high-tensile-strength fibers in the rim of the wheel to withstand the centrifugal forces experienced at high angular velocities. Figure 6.6 is a block diagram of a battery/flywheel system for an electric car in parallel configuration.[9]

*Figure 6.4. Layered disc flywheel prototypes. (F.J.M. Thoolen)*

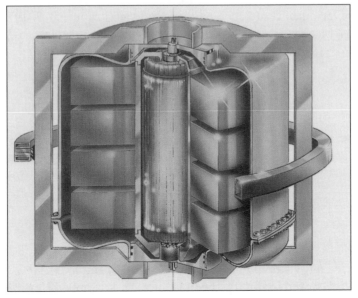

*Figure 6.5. An electrical-mechanical battery motor-generator system. (Ref. 8. F.J.M. Thoolen. Cover: Ruud van der Hoorn/Roy Mohede-CCM, Nuen. Teo van Gerwen-Design, Leende.)*

123

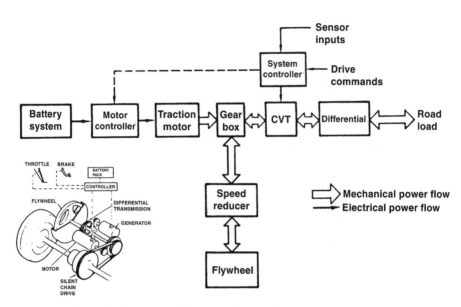

*Figure 6.6. Simplified block diagram of battery/flywheel propulsion system.*
*(Lawrence Livermore National Laboratory)*

With the Electric and Hybrid Vehicle Act of 1976, the dual-powered automobile was slated to be investigated and the foci found for the most productive effort to determine the viability of the flywheel as an electric vehicle range extender. Menker[3] relates that the flywheel as an energy storage device, despite its early historical use, is a concept which is only in its infancy—the first technical conference on the subject was as recent as 1975. On a stored energy/mass basis, Menker gives the specific energy for several devices in Table 6.1.[3]

### TABLE 6.1
### ENERGY STORAGE COMPARISONS

| | |
|---|---|
| Rubber bands (Carli's 1894 multi-powered tricycle) | 9 watt-hours/kg |
| Lead-acid battery | 18–33 watt-hours/kg |
| Flywheel (advanced anistropic material) | 190–870 watt-hours/kg |
| Gasoline | 13,300 watt-hours/kg |

Source: Ref. 3

Note that the least energy stored in Menker's flywheel example is 190 watt-hours/kg. In contrast, the energy stored in the flywheel used by IKA is only 175 watt-hours/45 kg = 3.9 watt-hours/kg, a figure approximately 1/50 as much. With such a relatively high energy-to-mass ratio, the flywheel appears to merit investigation as an energy storage device, but ponder the same ratio for gasoline. Small wonder the world uses gasoline-powered vehicles. The energy contained in gasoline and the relatively simple internal combustion engine is indeed a remarkable combination. Comparing the Posts' article[4] and the computer simulation study on a flywheel-electric performed by Martin W. Schwartz,[10] the latter's conclusion appeared less sanguine for range enhancement.

## The AIResearch Manufacturing Company Flywheel-Electric Vehicle

On the basis of all the data and the Electric and Hybrid Vehicle Act, a competitive contract for design of a flywheel-electric vehicle was awarded to the AIResearch Manufacturing Company of California, a division of the Garrett Corporation, with the second phase being proof of concept construction of actual test vehicles.[11] Design work began in 1976.

The flywheel-electric vehicle is shown in Figure 6.7. A cutaway drawing is illustrated in Figure 6.8. The power system is a flywheel, traction motor, generator, gearbox, controller, and battery pack. As stated in Ref. 11:

- NEW POWER SYSTEM USES FLYWHEEL TO IMPROVE BATTERY UTILIZATION
- REGENERATIVE BRAKING RECOVERS ENERGY AND INCREASES RANGE
- ALL LIGHT WEIGHT FIBERGLASS CONSTRUCTION MEETS ALL FEDERAL SAFETY STANDARDS
- NEW HIGH ENERGY LONG LIFE BATTERIES
- CITY DRIVING RANGE . . . . . . . . 80 MILES
- CRUISING SPEED . . . . . . . . . . . . . 55 MPH
- ACCELERATION 0-55 . . . . . . . . . 20 SEC.

*Figure 6.7. A flywheel-electric automobile. (Garrett Corporation)*

*Figure 6.8. Interior of a flywheel-electric automobile. (Garrett Corporation)*

...the rotating shafts of the flywheel, generator, and traction motor are mechanically linked through the three power paths of the planetary gear-set and function together as an infinitely variable, electromechanical transmission. The flywheel shaft is connected to the sun gear of the differential planetary transmission. The generator is geared to the ring gear of the planetary, and the carrier output is connected to the final drive differential gearbox. The traction motor is geared to the carrier shaft.

Preliminary design specifications for the flywheel-electric vehicle are shown in Table 6.2.[11,12]

**TABLE 6.2**
**PRELIMINARY DESIGN SPECIFICATIONS**

| | | | |
|---|---|---|---|
| Overall Length, in. | 158 | Power System Type | Hybrid Electric |
| Width, in. | 70 | Controller Type | Ward Leonard |
| Height, in. | 57 | Cruise Speed | 55 mph |
| Wheelbase, in. | 95 | Passing Speed | 70 mph |
| Curb Weight, lb | 2566 | Speed, 5% Grade | 50 mph |
| Gross Weight, with Passengers, lb | 3166 | Range SAE J227D | 85 miles |
| Acquisition Cost, 1982 | $5000 | Range LA-4 | 94 miles |
| Ownership Cost/Mile | $0.15 | Range at 55 mph Cruise | 70 miles |
| Battery Weight, lb | 1040 | Range at 50 mph Cruise | 82 miles |
| | | Energy Use, City Driving | 0.391 kWh/mile |

This vehicle is unique in two major respects: 1) The power system includes a flywheel that stores energy during both charging and regenerative braking and makes possible the acceleration capability required to move with the traffic without reducing range to unacceptable values, and 2) Lightweight plastic materials are used to minimize weight and increase range, according to the developers.[13]

Another flywheel vehicle brought to fruition is shown in Figure 6.9, a U.S. Mail car. Of this mail car, Thomas A. Norman writes[14]:

> The flywheel system that has been designed and installed in the postal delivery vehicle replaces the old drive motor and control system with new units which are much lighter in weight. The result has been a reduction of 200 lb (91 kg) in the total weight of the vehicle. The flywheel system is designed to store the braking energy and use this energy to assist during acceleration and hill climbing. The result is an increase in acceleration, gradability, and number of starts and stops that can be completed. In addition, there is a reduction of battery peak current which should increase battery life and available energy.

Table 6.3 shows a comparison of the performance of the existing 1/4-ton electric vehicle and the flywheel system.

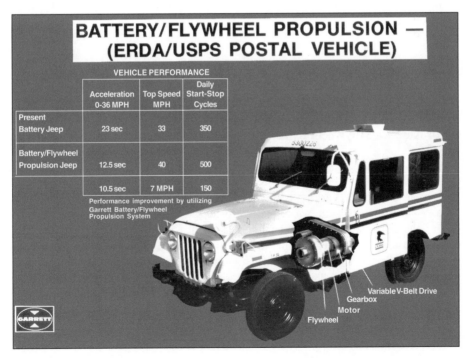

**BATTERY/FLYWHEEL PROPULSION — (ERDA/USPS POSTAL VEHICLE)**

VEHICLE PERFORMANCE

|  | Acceleration 0-36 MPH | Top Speed MPH | Daily Start-Stop Cycles |
|---|---|---|---|
| Present Battery Jeep | 23 sec | 33 | 350 |
| Battery/Flywheel Propulsion Jeep | 12.5 sec | 40 | 500 |
|  | 10.5 sec | 7 MPH | 150 |

Performance improvement by utilizing Garrett Battery/Flywheel Propulsion System

Variable V-Belt Drive
Gearbox
Motor
Flywheel

GARRETT

*Figure 6.9. A flywheel-electric equipped U.S. Mail car. (Ref. 14. Garrett Corporation)*

TABLE 6.3
ELECTRIC FLYWHEEL VEHICLE PERFORMANCE

|  | Present Electric | Present Hybrid |
|---|---|---|
| Acceleration (seconds) | | |
| 0 to 15 mph | 5.7 | 3 |
| 0 to 30 mph | 24.0 | 12 |
| Top Speed (mph) | | |
| Level | 32.5 | 40.0 |
| 5% grade | 20.5 | 25 |
| 10% grade | 14.0 | 20.0 |
| Range (Miles) | 30.0 | 30.0 |
| Simulated Route | | |
| Start/Stops | 300 | 500 |
| Miles | 8.5 | 14.4 |

In a preliminary design for an electric propulsion system using a flywheel, Francis C. Younger and Heinz Lackner studied 28 propulsion system concepts using various combinations of d-c and a-c motors. They concluded there is "no identifiable single best design to reach the desired range and performance objectives (for an electric vehicle), but rather a multiple of designs are possible."[15]

# American Flywheel Systems

Echoing the optimism of both Professors Post[4] and also of Menker[3] regarding the many cited advantages of the flywheel as an energy storage mechanism is the work of Edward W. Furia, president of American Flywheel Systems of Seattle, Washington. As reported by *Business Week*[16]:

> On June 23, American Flywheel Systems Inc. (AFS) in Seattle received a patent on a 'battery' it says will let cars get 600 miles (965 km) per charge—five times the current distance—without the toxic waste and corrosion problems of lead-acid models. The battery, charged from a 110-volt outlet, relies on kinetic energy. During charging, electrical current causes two (counter-rotating to minimize gyroidal effect) rotors, suspended by magnetic bearings in a vacuum, to spin at high speeds. Once the charge is completed, the rotors continue spinning. Twenty such batteries generate enough energy to drive a car-size motor for 43.6 kWh, vs. 13.6 kWh on General Motors Corporation's *Impact* prototype electric car.

Furia continues[17]:

We project that the AFS flywheel batteries, if installed in General Motor's two-passenger *Impact* electric car, would propel it between 300 to 600 miles (483 to 965 km) per charge, depending on the fiber composite used in the flywheel rotors. Twenty AFS flywheel batteries would fit into the present battery channel of the *Impact* and could store 43.6 kWh of energy—enough to drive the *Impact* 400 miles (644 km), roughly four times the 80- to 120-mile (129- to 193-km) range with the car's current 32 lead-acid batteries. If the *Impact* were reconfigured so that a total of 32 AFS flywheel batteries could be placed in the vehicle, it could go up to 600 miles (965 km) on one charge. Or, instead of increasing the range, flywheel batteries could be added to increase the payload, so that, for example, it might be possible for General Motors to build a viable four-passenger *Impact*.[c]

The *New York Times* of 3 January 1994,[18] in an article by Matthew L. Wald, relates that American Flywheel Systems will show a non-operating concept car at the Los Angeles Auto Show the following week, which is 9 January 1994. The basic shape of this car is the Chrysler *New Yorker*, without the engine, transmission, radiator, starter motor, gas tank, muffler, and tailpipe. With fiberglass substituted where feasible and with a flywheel battery and the electric motor, the resulting vehicle weighs 957 lb (434 kg) less than the gasoline version.

Although the stored mechanical energy in a flywheel varies as the square of its rotational speed, Furia hopes to have a rim speed for his flywheel, now being developed by Honeywell, to exceed the current record of 1400 meters per second (more than 3100 mph) now held by a flywheel developed at Oak Ridge National Laboratory. The present drive train with a 136-hp motor is manufactured by Albert Coccini of AC Propulsion Company. It was Coccini who developed the electronic control system for the permanent magnet, variable frequency synchronous motor for GM's *Sunraycer*. Such power, it is said, would give Furia's car an acceleration to 60 mph (97 km/h) in 6.5 seconds. Furia claims a range for his flywheel car, the AFS20, of 370 miles (595 km). Furia also believes the car could sell for $30,000. Figure 6.10 is an illustration of F.J.M. Thoolen and one of his special flywheels. Thoolen's Ph.D. thesis is cited in Ref. 8. Anyone who chooses to study the flywheel as an energy storage device should read this scholarly work.

# Continuing Work on Flywheels for Transportation

An excellent survey article on flywheels has been written by Steven Ashley of *Mechanical Engineering*. Ashley writes about 11 groups researching flywheels[19]:

---

[c] Using published data from General Motors Corporation, calculations indicate the originally designed *Impact* requires 113 watt-hours/mile while traveling at 55 mph (89 km/h) with the vehicle bearing 13.6 kWh of lead-acid batteries. Also stated: the range of the *Impact* is 120 miles (193 km) at 55 mph ( 89 km/h) at constant speed (or, according to Dr. Victor Wouk, 65 miles [105 km] in traffic). If Furia's flywheel batteries contained 43.6 kWh of energy (in *Impact's* same battery channel), the increased range factor for *Impact* would be 43.6 ÷ 13.6 = 3.2, or a range of approximately 3.2 × 120 = 390 miles. The relative cost of the flywheel batteries versus the lead-acid batteries is not given. Lead-acid batteries in the United States, now made in large quantities, are approximately $1 per pound (0.45 kg), or approximately $1 for 16 watt-hours in 1997.

*Figure 6.10.  F.J.M. Thoolen with one of his specially wound flywheels.  (F.J.M. Thoolen)*

- American Flywheel Systems (AFS) of Medina, Washington, has garnered a great deal of publicity for patenting a dual-rotor counter-rotating flywheel energy storage (FES) system.  The flywheel design, however, will not necessarily be used in a demonstration FES system being developed with Honeywell Satellite Systems of Phoenix, Arizona, that is due to be ready for testing in May 1994.  Honeywell, which has built 90% of the altitude-control gyroscopes and associated control electronics used in the U.S. space program, is investing $200 million in expertise and technology—power and control electronics, magnetic bearings, vacuum containers, and motor/generators—in the small flywheel company in exchange for a license on the AFS technology and any technology developed by the joint venture applicable to aerospace uses.

  AFS chairman and chief executive officer Edward W. Furia said that the joint venture is now testing a proof-of-concept FES system—a space reaction wheel system modified for terrestrial use.  Plans call for the construction of an FES powered vehicle by July 1995.  Furia added that AFS recently received a $2 million grant from the U.S. Advanced Research Projects Agency, the Sacramento Municipal Utility District, and a consortium of California state agencies.

- Advanced Controls Technology Inc. (Avcon) of Northridge, California, is thought by some researchers to have a lead in magnetic bearings for FES applications.  "We have a magnetic bearing on line in the lab that is specially designed for flywheel applications," said Crawford Meeks, Avcon president and chief executive officer.  "This hybrid permanent magnet/electromagnet bearing design was developed to keep electrical and rotating losses to absolute minimum," he said. Normally, the permanent magnets keep the bearing in metastable equilibrium so that it draws virtually zero power, Meeks explained. The electro-magnetic coil engages only when there are perturbations in the bearing's balance.

"There is a widespread misconception that magnetic bearings have no rotating losses," Meeks said. "But they do develop eddy current and hysteresis losses when they run at high speed." To combat these phenomena, Avcon engineers designed the low-loss homopolar-type bearing to minimize changes in the frequency and the amplitude of the magnetization field in the rotating part.

Negotiations on three flywheel programs—one with a private U.S. investor, one with a European company, and another with Calstart, a regional defense technology-conver-sion consortium—are on the verge of being finalized, Meeks said. One of these projects reportedly concerns FES magnetic bearing technology for railroad trains. Finally, Avcon is involved in a strategic partnership with part-owner AlliedSignal Inc. in Morristown, New Jersey, that aims to develop FES "power-generating stations."

- During the mid-1980s, a significant amount of research on the design and manufacture of prototype composite flywheel rotors was supported by Canada's now inactive National Energy Program. The technical know-how developed then by Ralph Flanagan and his team has been transferred to a newly formed company called Flywheel Energy Systems in Ottawa, Ontario, which has been collaborating with Thortek Inc. of Knoxville, Ten-nessee, and Mechanical Technology Inc. (MTI) of Latham, New York. The company is looking to promote Canadian participation in FES development.

Flanagan's most successful rotor is a 'bi-annular-type' design with an aluminum 'flex-rim' hub attached to two thick, concentrically mounted composite rings. Fabricated by wet filament winding, an inner S2-glass fiber composite ring and an outer carbon fiber composite ring are thermally assembled to the hub and each other by cooling the inner members with liquid nitrogen and then allowing them to expand into place inside the outer member to form an interference fit. This arrangement preloads the composite materials with compressive stresses, which partially cancel the tensile stresses pro-duced when the rotor spins at high speeds. The group is currently designing a new generation of rotors capable of 90,000 rpm (de-rated), depending on the diameter.

- Richard Post, a senior scientist, and his associates at Lawrence Livermore National Laboratory (LLNL) in Livermore, California, who have built and tested small experimen-tal FES systems, are now "close to running a completed flywheel backup power system for the laboratory's computer center." He noted that the laboratory is negotiating industrial corporate research and development agreements for this stationary applica-tion, which is conceptually similar to a vehicular system.

The main feature of the 3-kWh FES unit, which is the size of a small filing cabinet, is a 10-inch (25-cm) diameter high rotor composed of a nested set of thin-walled concentric cylinders fabricated of filament-wound unidirectional high-strength fibers embedded in an epoxy matrix.

The design avoids the delamination problem while at the same time achieving maximum strength in hoop tension. These cylinders are coupled together by compliant mechanical elements that transmit torques but not the strong radial forces. The thin cylinders ensure that centrifugally induced radial stresses remain within the composite's strength limits.

Figure 6.11 shows Dr. Richard F. Post and his equipment.

*Figure 6.11. Dr. Richard F. Post, veteran flywheel researcher and senior scientist at Lawrence Livermore National Laboratory, displays a carbon-fiber composite rotor cylinder (right) that is one of a nested set of concentric rotor shells for a 3-kWh flywheel energy storage system. This flywheel is being developed as a backup power source for the laboratory's computer center. On the left is a model used in developing the system's magnetic bearings.*
*(Lawrence Livermore National Laboratory)*

Post reported that his team is trying to develop a simple, totally passive bearing with no control circuitry.

Energy is coupled in and out of the rotor via a compact 'ironless' generator-motor, the magnetic field of which is produced by a special array of permanent magnets that concentrate a very uniform magnetic field in the interior while practically canceling outside fields. This (assembly) is the Halbach array, which was developed by researcher Klaus Halbach of Lawrence Berkeley National Laboratory in Berkeley, California, for accelerator applications.

• Specializing in high-speed rotating machinery, MTI is pursuing an ongoing internal project to demonstrate a 0.5-kWh FES system that uses a 15-inch (38-cm) diameter fiber composite rotor designed by Ralph Flanagan that rotates on a magnetic bearing at 30,000 rpm, said Paul Lewis, manager of core technology at MTI.

• Researchers at Oak Ridge National Laboratory (ORNL) in Oak Ridge, Tennessee, have extensive experience with developing composite rotors and high-specific-power axial-gap electric motors/generators. In the late 1980s, ORNL scientists constructed an experimental FES system for short-term high-power discharging for aerospace or defense purposes.

John Coyner, program manager for flywheel and composite technology at ORNL, predicted that flywheel vehicles will be demonstrated by 1995, and limited commercialization would begin in the 1996–1997 time frame.

• SatCon Technology Corporation, a spin-off from MIT's Draper Laboratories in Cambridge, Massachusetts, has designed and built many FES systems, mostly for aerospace and defense application, said SatCon's David Eisenhaure. Little is known about the company's recent FES-related actions beyond a recently announced deal with Chrysler Corporation of Highland Park, Michigan, to develop "innovative new vehicle drive train components." SatCon is pursuing development projects on stationary FES with undisclosed electric utilities.

• Douglas G. Thorpe, president of startup company Thortek, is working to adapt existing FES components into a working demonstration (non-optimized) system with a rating of 1.33 kWh. Thorpe is using a 78-lb (35-kg) Flanagan rotor that is 2 ft (60 cm) in diameter and about 4.5 inches (11 cm) thick with a top speed of 22,900 rpm and magnetic bearings supplied by an unidentified manufacturer. It will be linked to a 25-hp motor. The integrated system will soon be tested at NASA's Johnson Space Flight Center in Houston. Thorpe said that NASA engineers are interested in connecting a flywheel system to a turbine to pump hydraulic fluid to move flight control surfaces on the fins of booster rockets.

• In the late 1980s, Flanagan's research led to a collaboration with Alcan International Ltd. in Cleveland, Ohio, and Unique Mobility in Golden, Colorado, to design key components of an FES system for load-leveling a chemical battery-powered electric vehicle. Unique Mobility developed a 30,000 rpm 40-kW motor/generator, but funds ran out before it was integrated with a Canadian rotor.

- James A. Kirk, a professor of mechanical engineering, and his colleagues at the University of Maryland in College Park have built and tested experimental components of FES systems. Among their earlier work is a 0.5-kWh, 55-lb (25-kg) system for satellite energy management designed under a NASA contract. Kirk said that his group is working on a 22-lb (10-kg) graphite/epoxy flywheel that has been tested at 20,000 rpm in a vacuum chamber. He noted that the rotor is capable of 45,000 to 90,000 rpm. The group is also developing a 20-kWh, 5-kW stationary utility load-leveling shaft-drive flywheel with Baltimore Gas & Electric Co.

- U.S. Flywheel Systems Inc. of Laguna Hills, California, led by chairman Bruce Swartout, working with French aerospace conglomerate *Aerospatiale* of Paris (plans) to build 14 flywheel units for satellite momentum control, which demonstrated feasibility of spinning a composite rotor at high speed on a magnetic suspension. One of the satellite units has operated reliably in orbit for 13 years.

Swartout said that U.S. Flywheel Systems, which is partially funded by Calstart, has patents on thick-rim rotor designs that control radial stresses by two methods. In one design, strong, light, stiff fibers are used in the outer portion of the rim, whose inner portion is fabricated from high-density lower-modulus fibers. In the other approach, the company uses a manufacturing process in which powdered metal is added to the composite materials in decreasing amounts as rotor radius increases. The finished rim therefore features a continuously decreasing density as rotor radius increases, which lowers radial stresses at speed. Other technology includes a one-piece hub design and a partially developed method to produce low-cost, high-strength silica fibers.

"We are looking at the first phase of making a 4-kWh mechanical battery prototype within the next year," Swartout said. "At that point, the idea is to manufacture 10 or 12 of them to propel an automobile as a demonstration within the next 18 months. We know that if we select a 9-inch (23-cm) wheel diameter, it would rotate at about 120,000 rpm."

In addition to these research efforts, entrepreneurial efforts were under way such as that of Harold and Benjamin Rosen of Rosen Motors in Woodmont Hills, California. In January 1997, they test drove a turbine-flywheel powered hybrid electric vehicle. Their flywheel was composed of a titanium hub and a high-strength, carbon-fiber composite cylinder capable of 60,000 rpm.[20] Their vehicle is described in greater detail in the next chapter.

# Summary Articles on Flywheel Energy Storage

As you can see from these examples, substantial work is being executed in flywheel motors/generators in America and abroad, presenting both theoretical and pragmatic arguments in this energy storage source for electric transportation, and in load-balancing for electric utilities:

- Post, Richard F., Fowler, T. Kenneth, and Post, Stephen F., "A High-Efficiency Electromechanical Battery," Proc. IEEE, Vol. 81, No. 3, March 1993. High-speed (200,000 rpm), magnetic suspension bearings, 1-kWh size.

- Thoolen, F.J.M., "Development of an Advanced High-Speed Flywheel Energy Storage System," Cip-Data Koninkluke Bibliotheek, Den Hag, Netherlands, pp. 197, 1993. Lower speed (17,000-rpm), ball bearings, built primarily for bus and train use.

Both papers have extensive bibliographies.

The next chapter discusses turbine drive, hybrid electric motorcars.

# Note 1—Another Flywheel System

Commonwealth Edison Company of Chicago and Argonne National Laboratory of Argonne, Illinois, are collaborating on developing a flywheel system as an electrical energy storage device to assist in resolving the perennial electric utility problem of load-leveling—to more nearly match daytime and nighttime electrical loads.[21] Although interesting, the subject is outside the purview of this book. See Note 4 for a summary of the Japanese experiments.

# Note 2—Gyroscopic Effects

Gyroscopic forces, in the case of a flywheel as an aid in electric utility load-balancing, are of little concern because the system probably would be operated in only a fixed position. However, in an automobile, where the flywheel is employed as an energy storage device, gyroidal forces must be considered. Steven Ashley, associate editor of *Mechanical Engineering,* writes on road shock and gyroscopic forces,[19]

> Two other concerns for designers of vehicular Flywheel Energy Storage (FES) systems are how to deal with dynamic loads from road shocks and how to handle gyroscopic forces encountered when a spinning rotor changes orientation in maneuvers. In general, transitory loads will be isolated from the rotor by shock-absorbing or elastomeric systems.

This is shown in Figure 6.12.

> "The gyroscopic effect is dealt with more easily with small flywheels modules," (Professor Richard F.) Post said, "because the scaling of the rotor's angular momentum is proportional to the fourth power of its radius, meaning that small rotors will exert markedly smaller gyroscopic effects than larger ones.[22] The gyroscopic moment of a 1-kWh flywheel at maximum speed is comparable to that of a typical auto engine flywheel at maximum rotation," he said. Zorzi, who is Vice President of Engineering and Technology at American Flywheel Systems, concurred, saying "You must pay attention to the gyroscopic effect in maneuvering conditions because it is significant, but it's not disabling."

> Under consideration are gimbal mountings that allow the rotor to maintain orientation as the vehicle moves, counter-rotating rotors that inherently counteract gyroscopic forces, or the use of an even number of FES units, which would ameliorate much of the gyroscopic effect. It should be noted that whatever scheme is used, the magnetic bearings must be able to provide the necessary restraining torques to handle inputs not damped out by shock- and vibration-isolation systems.

"In addition to controlling gyroscopic effects, the use of small flywheel modules provides other advantages," Post noted. "It provides very high power densities and reduced gyroscopic moments; it makes it easier to use rotors in pairs and to contain failed rotors. The use of a modular approach also provides redundancy and simplifies manufacturing and transportability."

Given the current state of the technology (Zorzi said), "the challenge is to match the components to the system and successfully integrate them." By the time the technology is available, however, FES may face tough competition from advanced chemical batteries, fuel cells, super capacitors, and other energy-storage devices currently in development.

"In the end, factors other than technology will determine whether flywheels prove successful," said George Chang, chief of the mechanical energy branch of the U.S. Department of Energy (DOE) Office of Energy Technology in the late 1970s and now a professor

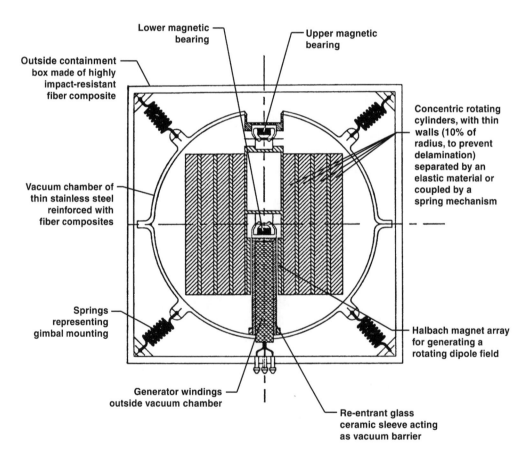

*Figure 6.12. A cutaway view of the backup-power flywheel battery for the computer center at Lawrence Livermore National Laboratory, which shows how the multi-ring carbon-fiber composite rotor spins almost friction-free on a magnetic bearing while suspended in an evacuated chamber. (Ref. 19. Steven Ashley)*

of aerospace engineering at Embry Riddle Aeronautical University in Daytona Beach, Florida. "It's really not a matter of technology," he said. "It's an institutional question because it will take strong incentives to switch away from current technology and the industries that have been established to support it. In addition, it will take a major effort to understand the economics of flywheel-powered EVs and to answer questions about the infrastructure needed to make the technology work."

Another application of using a flywheel as an energy storage device has been devised by J.P.M. Parry & Associates Ltd., a subject discussed in Note 3.

# Note 3—The Parry People Mover

**Vehicle Drive System.** J.P.M. Parry & Associates Ltd. of West Midlands, United Kingdom, offers a flywheel-based tram that has characteristics similar to the flywheel bus shown in Figure 6.1. Differences are primarily on the means of energy feed and the means of road contact. The Parry People Mover (PPM), shown in Figure 6.13, is described by the manufacturer as having the energy stored "in the vehicle by a large rotating flywheel. Drive to the wheels is obtained by linking this (flywheel) through clutches and a continuously variable transmission (CVT), which gives control over acceleration and cruising speed. The transmission is also used to slow the vehicle to crawling speed and also to limit downhill speeds. This control is achieved by 'regenerative braking' whereby the revolutions per minute (rpm) of the flywheel are increased when the vehicle is slowed." The transmission system is based upon conventional technology and, it is claimed, is extremely straightforward to service and maintain.

**Charging Points/Boarding Platform.** "After travel between stations, the flywheel rpm will have decreased slightly. Energy is transferred to the flywheel from an external electrical connection, which provides power for an onboard 70-volt traction motor. (See Figure 6.14.) The flywheel is wound up for only 20 to 30 seconds, which is sufficient to bring the rpm up to the desired level, during which time passengers will be boarding the vehicle." (The company does not treat operations under snow conditions).

"The charging points are constructed with a shrouded charge rail that, if necessary, would become live only when the vehicle is alongside. (See Figure 6.13.) The boarding platform is built over the charge rail at the same height as the vehicle floor, allowing easy access for wheelchairs, prams, etc."

**Tracking Technology.** "Due to the use of individual charging points at the boarding stops, the PPM system does not require electrified track or overhead wires. Simple, lightweight track can be used based on either special tramway section steel or typical light rail. Thirty pounds per yard rail is sufficient for the heavier PPM vehicles. The track does not require extensive foundation work, and, in the case of street running application, the rail can often be laid in the existing road surface. In some cases, the use of paving blocks will enhance the appearance of the tramway."

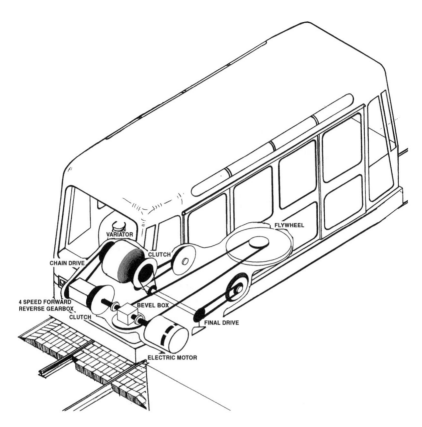

*Figure 6.13. Vehicle drive system. (Parry People Movers, Ltd.)*

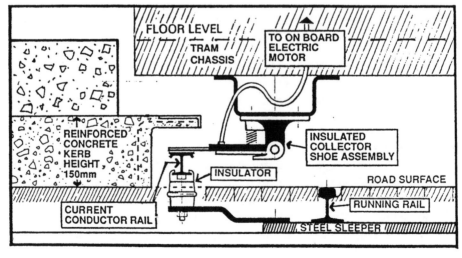

*Figure 6.14. Details of charging point. (Parry People Movers, Ltd.)*

# Note 4—Flywheel Energy Storage in Japan

According to the Japanese group cited in Ref. 23:

> Research progress in the research and development of yttrium-based oxide high-temperature superconductor has enabled the production of a large-diameter bulk (mass) with a strong flux-pinning force. A combination of this superconductor and a permanent magnet makes it feasible to fabricate a noncontact, noncontrolled superconducting magnetic bearing with a very small rotational loss. Use of the superconducting magnetic bearing for a flywheel energy storage system may pave the way to the development of a new energy storage system that has great energy storage efficiency. From relevant data measured with a miniature model of the high-temperature superconducting magnetic bearing, we developed a conceptual design of an 8 MWh flywheel energy storage system using the new bearing, which proved to be potentially capable of achieving a high energy storage efficiency of 84%. Based on the results of these research activities, we built a 100 Wh-class experimental system that attained a high revolution rate of 17,000 rpm. In addition, we found that magnetic bearing using the high-temperature superconductor was subject to only a very small rotational loss of around 0.6 watt.

Assuming a daily operating pattern of four hours of energy input and four hours of energy output (quite different for a vehicle regenerative braking system), the Japanese calculated a daily energy storage efficiency to be approximately 84%. For a battery energy storage system, they give 70 to 73%.

# References

1.  "The Oerlikon Electrogyro: Its Development & Application to Omnibus Service," *Auto. Eng.*, December 1955.
2.  *Engineering News*, **102**, August 1993, p. 10, United Kingdom. See also: Vaughn, Mark, "Reinventing the Wheel," *AutoWeek,* 3 March 1997, pp. 16–17.
3.  Menker, Sherwood B., *Flywheel Energy Storage, Mark's Standard Handbook of Mechanical Engineers*, 8th ed., McGraw-Hill, New York, 1978, pp. 9–170, 9–171. See also: *McGraw-Hill Encyclopedia of Science and Technology*, Vol. 7, 1992, pp. 232–234.
4.  Post, Richard F., and Post, Stephen F., "Flywheels," *Scientific American,* **229**, 6 December 1973, pp. 17–23. See also: Post, Richard F., Bender, Donald A., and Merritt, Bernard T., "Electromechanical Battery Program at the Lawrence Livermore National Laboratory," AIAA-94-4083-CP, American Institute of Aeronautics and Astronautics, 7–11 August 1994.
5.  Wakefield, Ernest H., *History of the Electric Automobile: Battery-Only Powered Cars*, Society of Automotive Engineers, Warrendale, PA, 1994.
6.  Stephenson, R. Rhoads, *et al.*, "Should We Have a New Engine?" Technical Report, JPL SP 43-17, Vol. II, Jet Propulsion Laboratory, California Institute of Technology, Pasadena, CA, August 1975, Chapter 9, p.5.

7.  Helling, J., *et al.,* "Hybrid Drive with Flywheel Component for Economic and Dynamic Operation," Institut fur Kraftfahrwesen Technische Hochschule Aachen, Proc. Third International Electric Vehicle Symposium, February 1974.

8.  Thoolen, F.J.M., "Development of an Advanced High-Speed Flywheel Energy Storage System" (Ph.D. Thesis, 1993), Horn, Netherlands. See also: Barlow, J.M., *et al.,* "Mechanical Energy Storage Technology Project, Annual Report for Calendar Year 1979," Lawrence Livermore National Laboratory, Livermore, CA, 1 May 1980, p. 137.

9.  O'Connell, L.G., *et al.,* "Energy Storage Systems for Automobile Propulsion: Final Report, V. 1, Overview and Findings," Lawrence Livermore National Laboratory, Livermore, CA, 15 December 1980, p. 19.

10. Schwartz, Martin W., "Assessment of the Applicability of Mechanical Energy Storage Devices to Electric and Hybrid Vehicles," V. 2, Jet Propulsion Laboratory, California Institute of Technology, Pasadena, CA, 1 May 1979.

11. "Flywheel to Improve EV Performance," *Electric Vehicle News*, August 1977, p. 13.

12. "Test and Evaluation of the DOE ETV-2 Flywheel Electric Vehicle," 1982 SAE International Congress, 22–26 February 1982, Detroit, MI, Society of Automotive Engineers, Warrendale, PA, 1982.

13. "Near-Term Electric Test Vehicle ETV-2," AIResearch Manufacturing Co., Los Angeles, CA, Contract AC03-76CS51213, TIC Order No. DE83006423, Reference DOE/CS/51213-01 (WHK).

14. Norman, Thomas A., "Electric-Flywheel Vehicle for Postal Service Applications," SAE Paper No. 780297, Society of Automotive Engineers, Warrendale, PA, 1978.

15. Younger, Francis C., and Lackner, Heinz, "Study of Advanced Electric Propulsion System Concept Using a Flywheel for Electric Vehicles," DOE/NASA/0078-79/1, National Aeronautics and Space Administration, Lewis Research Center, Cleveland, OH, December 1979.

16. Buderi, Robert, *Business Week*, 13 July 1992, pp. 139.

17. Furia, Edward W., *Business Week*, 13 July 1992, pp. 10–11.

18. Wald, Matthew L., *New York Times*, 3 January 1994, p. C21.

19. Ashley, Steven, "Flywheels Put a New Spin on Electric Vehicles," *Mechanical Engineering*, October 1993, pp. 44–51.

20. Rosen, Harold A., and Castleman, Deborah R., "Flywheels in Hybrid Vehicles," *Scientific American*, **227**, October 1997, pp. 75–77.

21. Van, John, *Chicago Tribune*, 19 September 1993, Sec. 7, p. 10.22. Otaki, Hideyuki, "Some Analysis of Gyroidal Effect and the Development of a Flywheel Powered Vehicle," SAE Paper No. 800835, Passenger Car Meeting, 9–12 June 1980, Society of Automotive Engineers, Warrendale, PA, 1980.

23. Shinagawa, Jiro, Shikoku Electric Company, and Ishikawa, Fumihiko, Shikoku Research Institute, Japan, "Research and Development Project for Flywheel Energy Storage System Using High-Temperature Superconducting Magnetic Bearing," New Electric 21: Designing a Sustainable Electric System for the Twenty-First Century, Conference Proceedings, Paris, France, 22–24 May 1995.

# CHAPTER 7

# Gas Turbine-Electric Motorcars

Joining the panoply of possible solutions in overcoming range limitation of the electric car is the use of gas turbines as a supplementary source of power. This chapter limits discussion to: 1) a short history of the gas turbine, 2) the Chrysler Corporation concept car, 3) the Volvo hybrid electric car, and 4) the Rosen Motors turbine-flywheel power train. Other designs are in the wings. However, what was the background in developing this prime mover? The gas turbine has a surprisingly early history.

## History of the Gas Turbine

France's Montgolfier brothers, Joseph and Jacques, in 1782 observed that a hot-air-filled balloon would ascend to higher elevations, even while bearing a duck, a rooster, and a sheep. In the same way, an unknown home experimenter assembled the first primitive, hot-air, gas turbine. The latter machine had initial use in turning a spit for roasting meat. In effect, propeller blades were placed on a suspended vertical shaft in the chimney. Hot air rising from the blaze in the fireplace rotated the blades. Suitable rods and gearing caused the spit to turn slowly, evenly roasting the meat. As early as 1791, John Barber of England was granted a patent for a gas turbine. However, the first gas turbine with an accompanying compressor was created in France in 1894 by R. Armengand and C. Lamale. Continuing, F. Stolze of Germany in 1900, Stanford A. Mass of the United States in 1902, and A. Elling of Norway in 1903 made further contributions to the theory of gas turbines.

By 1939, as a result of progress in aerodynamics during the 1930s, sufficient advances had been made in gas turbine theory to encourage the Brown-Boveri Company of Switzerland to build a gas turbine connected to a 4000-kW electric generator. During this time, immediately before World War II, Frank Whittle in England and Hans-Joachim Pabst von Ohain in Germany considered gas turbines for what were soon to be called jet aircraft. Those readers of sufficient age will recall the dire impact of these first German jet fighters, which clearly outflew the best propeller planes airborne by the Allies at the time. However, these Nazi attack fighters appeared so late in the war that they had little influence on its outcome.

In America, gas turbine work continued on the Whittle design, and this power system now is used almost universally on large aircraft. Other uses for gas turbines, in conjunction with a generator, are as a standby power source for electric utilities to handle peak electric loads, in some high-speed naval ships, and as a power source for pumping fluids through pipelines.

Several of the major car makers have made concept cars with gas turbines. In 1963, Chrysler built a fleet of 50 gas turbine cars. For more on the Whittle jet engine and Chrysler's gas turbine cars, see Notes 1 and 2 at the end of this chapter.

## Chrysler and Volvo Motorcars

Both the Chrysler and Volvo automobiles use turbines to drive an a-c generator.[1] With the Chrysler approach, the generator, an integral part of the flywheel, is on the same shaft as the turbine, the former serving as an energy reservoir. In the Volvo design, the turbine and generator are also on the same shaft, but a nickel-cadmium battery serves as the energy storage element. The traction motor of the Chrysler car is a three-phase, squirrel-cage, induction motor, as is the Volvo car.[2] The speed of rotation of an induction motor is frequency and number of poles dependent, less the effect of load. This difference in speed of rotation of the rotor, relative to the angular velocity of the rotating electromagnetic field established by windings in the stator, is known as *slip*.

The drive motor could be a permanent magnet, synchronous motor. In normal operation, the synchronous motor is only frequency and number of poles dependent. There is no slip. Because of the substantial cost of the permanent magnet, the induction motor is considerably less expensive than the synchronous motor. However, the relative efficiencies of the two motors/generators are almost the same. The synchronous motor is approximately 1 to 2% more efficient, but both are near the 95% range, depending on size and load.

In both cars, the turbines are run at constant speed with minimum emissions, the optimum employment for a turbine. In the Chrysler motorcar, the flywheel absorbs or supplies (balances) energy from the constant feed from the turbine, and the varying demands of the a-c traction motor. In the Volvo design, a nickel-cadmium battery furnishes a similar function. With this short introduction, we can now discuss both vehicles.

# The Chrysler Turbine-Flywheel Car—The *Patriot*

Periodically, transportation companies return to the age-old concept for storage of energy—the flywheel. For example, Lockheed determined that the power density of a steel flywheel rotor could exceed a specific power of 5000 watts/lb (11 kW/kg),[3] and Applied Physics Laboratory obtained 28 watt-hours/lb (62 watt-hours/kg) rotor energy density (specific energy) with bar-type filamentary flywheels.[4] Comparable figures for a top lead-acid battery would be 45 and 17, respectively.[5] Those users interested in hybrid electric vehicles found it refreshing to learn that Chrysler recently announced a race car bearing a turbine-flywheel electric drive system.[6]

Of this car, Matthew L. Wald writes,[7]

> The combination of the (clean burning) turbine and the flywheel provides 500 hp for extremely quick acceleration—the car could reach 200 mph (322 km/h)—and the flywheel recaptures the energy lost in braking. The flywheel has a built-in device that can be either a motor or an electric generator. When excess energy (from the turbine) is available—for example, when the car is not running at top speed—the motor spins the flywheel faster. When more energy is needed by the car than the turbine can provide—when it is accelerating, for example—the flywheel motor reverses function. It takes mechanical energy from the flywheel and converts it into electric energy to turn the wheels.
>
> This turbine is like a jet engine, only this one will turn faster than a typical jet. Turbines use fuel efficiently, but they are not good at varying their speed in the way cars typically need.
>
> François J. Castaing, vice president of vehicle engineering at Chrysler, said that the car, called the *Patriot*, could form the basis for ordinary passenger cars. Disk brakes and fuel injection, two technologies in near-universal use today, both began on the racetrack, he said.
>
> Although this flywheel is intended for use as part of a hybrid (automobile), others are studying flywheels as the energy storage devices for all-electric cars. The *Patriot* would not help Chrysler meet the California car rules, which will require battery-only electric cars beginning in 1998. But in broad outline, it is one of the kinds of cars envisioned by President Clinton and Vice President Al Gore in the 'super car program,' a partnership between auto manufacturers and the nuclear weapons laboratories that was announced at the White House (in the fall of 1993)...
>
> David B. Eisenhaure, president of the Satcon Technology Corporation of Cambridge, Massachusetts, which is building the flywheel and the generator that would make electricity from the turbine, said that for a passenger car, the whole drive train, including turbine, flywheel, and motor for the wheels, could weigh as little as 200 lb (90 kg). The turbine could be quite small, he said, because the flywheel would store energy for quick starts. The wheel, a 20-inch (51-cm) disk made of carbon fibers, would spin at the edge at nearly 1400 meters a second, about 3000 mph (4830 km/h). Most electric cars can use their drive motors as generators to capture energy when they brake, but ordinary lead-acid batteries

cannot absorb the power very fast. (This limitation is discussed in detail in Ref. 8.) The flywheel, however, can take in most of the power. In the *Patriot,* Mr. Eisenhaure said, the flywheel's regenerative braking will extend the vehicle's range 30%.[a]

Figure 7.1 shows the Chrysler car.[9]

# The Volvo Gas Turbine Electric Car

Regarding the Volvo gas turbine electric car, Michael Valenti, associate editor of *Mechanical Engineering,* writes,[10]

> Swedish engineers have designed a specialized electrical generator, gas turbine, electrical-drive system and automated vehicle-control system to reduce automotive emissions without compromising energy efficiency. These components have been designed into a four-door sedan, called the Volvo Environmental Concept Car (ECC), that uses its gas turbine for highway driving and to recharge the battery that powers its electrical drive system in the city. In addition to running on either fuel or electricity, the Volvo ECC is a true hybrid-drive vehicle that can shift to turbine or battery power by means of a programmable vehicle-management unit. While the Volvo ECC is not yet commercially available, its design and ongoing testing are providing valuable emissions and performance information for making hybrid-drive cars a reality.

> **Government and Business Join Forces**

> Work on the Volvo ECC began in 1986 when the Royal Institute of Technology in Stockholm, Sweden, began a research project to develop a high-speed generator with an electrical output of 20 kW at 100,000 rpm that could be used together with a gas turbine in a hybrid drive automobile. The project was funded by the Swedish branch of the multinational engineering firm Asea Brown Boveri (ABB) Ltd. in Zurich, by the Swedish power company Vattenfall A.B. in Vallingby, by automaker Volvo A.B. in Göteborg, and by the Swedish National Board for Industrial and Technical Development in Stockholm.

> In 1990, ABB, Vattenfall, and Volvo joined forces to build an experimental hybrid car featuring the high-speed generator (as shown in Figure 7.2). The vehicle that emerged, the Volvo ECC, is equipped with a gas turbine that powers a permanent-magnet synchronous high-speed generator to provide the electricity for a three-phase induction motor and a nickel-cadmium storage battery.

> The Volvo ECC is operable in one of three modes. The driver can select the electric-drive mode for short runs on city streets, in which the car runs solely on its battery; the combustion engine mode, for long drives on the highway; or the hybrid-drive mode, where the automobile's gas turbine cuts in when the battery charge drops below the 20% mark or when there is not enough charge available for a task such as acceleration. A small display

---

[a] I believe this number probably is high by a factor of two, depending on driving conditions.

*Figure 7.1. The Chrysler turbine-driven twin alternators weigh 190 lb (86 kg) and generate an a-c current that switches between the induction motor, capable of speeds up to 24,000 rpm, and the flywheel. The gimbaled flywheel weighs 147 lb (67 kg) and spins in a vacuum on magnetic bearings at speeds up to 58,000 rpm. (Ref. 9. Chrysler Corporation)*

*Figure 7.2. Outwardly similar to conventional four-door sedans, this Volvo concept car combines electrical batteries and a gas turbine to minimize emissions without reducing performance. (Environmental Competence Center, Volvo Car Corporation)*

in the battery ammeter shows the driver the level of battery charge when the car is in its electric-drive mode. A tachograph tells the driver how much mileage is left in the car for both the electric- and hybrid-drive modes.

### Making a Low-NO$_x$ Turbine

A key consideration for the Volvo ECC designers was the construction of a low-emitting gas turbine. The natural gas turbine in the Volvo ECC was developed by Volvo Aero Turbine, formerly United Turbine, a subsidiary of Volvo. Through another subsidiary, Volvo Flygmotor, the automaker has developed and produced jet engines for the Royal Swedish Air Force.

The Volvo ECC is equipped with a single-stage centrifugal compressor made of aluminum; the turbine itself is a temperature-resistant nickel alloy. Fuel and air entering the turbine's combustion chamber are intimately mixed and vaporized prior to combustion. 'This lowers the flame temperature and minimizes the formation of nitrogen oxide,' said Johnny Rehn, a design engineer at Volvo Aero Turbine who developed the Volvo ECC gas turbine. Rehn equipped the gas turbine with a regenerative, or rotary, air heater that serves as a heat exchanger. This unit consists of a ceramic rotor perforated with numerous passages through which hot gases and cool air flow alternately. Heat from the gases is absorbed by the air during the second half of each turbine revolution to improve turbine efficiency. This resembles the principle of recapturing waste heat that is used in cogeneration to boost power-plant efficiency. The Volvo hybrid-drive gas turbines tested thus far have been fueled by diesel oil, but the turbines can also burn gasoline, methanol, and natural gas, Rehn said.

*Figure 7.3. The gas turbine and the high-speed generator of the Volvo ECC share the same shaft so that the generator can be powered by the turbine. (Environmental Competence Center, Volvo Car Corporation)*

### Generator Maximizes Magnetic Circuit

The shaft of the gas turbine is shared by the synchronous high-speed generator of the Volvo ECC so that the generator is directly driven by the turbine. Both the compressor and the turbine are self-supporting units so that a single bearing on either side of the generator rotor is sufficient. Using a single shaft and a single turbine simplifies the assembly and reduces its costs, Rehn said. (See Figure 7.3.)

The Volvo ECC generator is excited by a permanent neodymium-iron-boron magnet. The generator was designed to work with gas turbines that have an output of 38 kW at 90,000 rpm. This eliminates the need for shifting into a lower gear which is typically required by the high rotational speeds of automotive gas turbines. The rotor enclosure was designed of high-strength steel to counteract the strong centrifugal forces generated by the high rotational speeds. These forces impose limits on the rotor diameter, which is a problem because the torque is proportional to the rotor volume in electrical machines. Additionally, the critical speed limits the practical length of the rotor. Thus, the challenge facing the high-speed generator research team was to achieve stable magnetization in spite of the small rotor volume, while ensuring sufficient magnetic flux in the magnetic circuit consisting of the generator's rotor and stator.

To minimize the rotor and stator losses that result from stator teeth, engineers used an air-gap wound stator. The permanent magnet, stator core, and air gap serve as the magnetic circuit, providing a number of benefits, including maximum magnetic volume, optimum use of the permanent magnet and the enclosure material, sinusoidal magnetic flux, low surface losses, low reactances, and minimal risk of demagnetization. The project engineers provided the high-speed generator with a ring winding because it is easier to install than a drum winding. Ring windings increase overall machine diameter, but the (heat generated by the) stray losses they cause in the stator casing are dissipated by cooling the unit with compressed air. Litz-wire stranded conductors with insulated filaments are used to help

147

keep the current from running along the sides of a solid conductor. (Due to different internal inductances, the conductor carries a high-frequency current). The winding material also prevents the high stray losses that would otherwise result from the main flux penetrating the winding. The winding is cast in epoxy resin containing boron nitride to increase its thermal conductivity.

### Enhancing Motor Performance

The high-speed induction motor of the Volvo ECC requires a high-speed reduction ratio of 10 to 1. Volvo engineers designed a special two-speed automatic transmission to retain the benefits of the electric drive while avoiding the use of a manual gearbox.

The new transmission works together with a transmission controller. The controller measures the revolutions per minute at the drive-end and output and controls the gear shifting by changing its hydraulic pressure. The vehicle-management unit communicates with the transmission and adjusts the engine torque accordingly. This ensures smooth gear changing without a torque converter. No reverse gear is necessary because the direction of rotation of the three-phase motor can be reversed by changing its phase sequence.

The transmission is attached to the Volvo ECC electric drive motor, whose design was based on a standard three-phase M13T 132-4 cage induction motor that delivers a nominal rating of 7.5 kW at 1500 rpm. ABB design engineers increased the motor speed, added water cooling, and used a stator winding with a higher fill factor to enhance the concept car's motor performance. The motor (shown in Figure 7.4) can develop 70 kW at 12,000 rpm with a maximum torque of 180 Newton-meters.

*Figure 7.4. A synchronous generator, gas turbine, three-phase motor, and transmission are mounted on the front axle of the Volvo ECC. (Environmental Competence Center, Volvo Car Corporation)*

The electric-drive motor of the Volvo ECC is powered by a conventional nickel-cadmium battery with a high power-to-weight ratio. While the battery is currently too expensive to be used commercially, less-costly lead-acid batteries would limit the vehicle to moderate acceleration and operating ranges. The battery of the Volvo ECC has two functions. In the hybrid-drive mode, the battery serves as an energy storage unit, allowing the car's gas turbine to operate at optimum efficiency under all conditions. The Volvo ECC's fuel consumption is 5.2 liters per 100 km, or 45.2 mpg, on the highway, and 6 liters per 100 km, or 39.2 mpg, on city streets, provided no net power is taken from the battery by other functions of the car, such as lights or air conditioning.

In the electric mode, the battery is the sole energy source for the Volvo ECC. The battery gives the car a range of 85 km, or 53 miles, in urban traffic.

The gas turbine of the Volvo ECC is used to recharge the battery during low-load periods, when the power needed for driving is less than what the turbine produces, or during regenerative braking, that is, when braking in the battery mode. Recharging is controlled by the car's vehicle-management unit and is based on how far the battery charge has dropped from the ideal condition, which is about 80% of full charge. This is because reserve storage capacity must be available to take advantage of the regenerative braking. The driver would only fully charge the storage battery prior to using the car for a completely battery-powered journey. Voltage from the battery is supplied to the motor by an inverter. Engineers used water-cooled isolated gate bipolar transistors to keep this unit compact. It measures 490 by 195 by 210 mm (19 by 7.6 by 8.3 inches). The inverter uses phasor-applied torque control based on the rotor speed feedback.

Hybrid-drive systems are typically plagued by electrical instability, which is commonly manifested in wide power swings between the drive motor and the synchronous generator. These oscillations often result from an imbalance between the energy input and output that causes the bus voltage to change. This voltage change leads to a load change in the motor, thereby causing a variation in the turbine speed. Fluctuations in speed increase the voltage change even more.

ABB engineers used a strategy employed in the energy industry to stabilize the electrical system of their hybrid drive Volvo. They connected the Volvo ECC battery to the electric drive train via a dc-to-dc converter in order to regulate the voltage. The dc-to-dc converter steps up the lower battery voltage, typically 120 volts, to the much higher bus voltage of 420 volts. This keeps the bus voltage constant even when the power load varies widely. An important consideration for the dc-to-dc converter was that it be compact. As in the case of the inverter, the engineers used water-cooled isolated gate bipolar transistors to keep the converter to the same dimensions as the inverter unit.

### The Vehicle-Management Unit

A programmable vehicle-management unit was designed in order to oversee the diesel oil, electric, and hybrid drives of the Volvo ECC. In addition to controlling the gas turbine and electric-drive systems, this computer controls vehicle start-up and shut-down; measures the battery charge; computes the car's range during electric and hybrid modes; displays information on the dashboard; monitors instrument, control, and pedal data; performs diagnostics; and records events. The vehicle-management unit also collects information for testing the vehicle's performance through a socket in the Volvo ECC's glove compartment. Technicians plug a laptop computer running specially designed data acquisition software into this socket to check variables, measured values, and control signals. This eliminates the need for conventional vehicle-testing instrumentation. In addition, the laptop can be used to change the system parameters and fine-tune the performance of the vehicle, monitoring the transmission for research purposes. The transmission is equipped with its own computer that communicates with the vehicle-management unit.

### Greener Pastures for Hybrid-Drive Cars

The development of cleaner-running automobiles like the Volvo Environmental Concept Car is underpinned by communities around the world seeking to reduce pollution. In May, Chicago became the 10th—and largest—city to join the U.S. Department of Energy's Clean Cities program, which aims to put 250,000 alternative-fueled vehicles on the road and between 500 and 1000 refueling stations in 50 cities around the country by 1996.

This demand for greener cars is not lost on the engineers testing and refining the Volvo ECC. It will be possible to build a hybrid-drive synchronous generator with lower emissions that can meet future EPA emission standards, according to Johnny Rehn of Volvo Aero Turbines. For example, the EPA emissions standards for light-duty vehicles of the 1996 model year will be 3.4 grams per mile of carbon monoxide, 1 gram per mile of nitrogen oxide, and 0.25 gram per mile of nonmethane hydrocarbons. In laboratory testing, the Volvo ECC produced 0.13 gram per mile of carbon monoxide, 0.17 gram per mile of nitrogen oxide, and 0.01 gram per mile of unburned hydrocarbons.

For a more detailed description of the Volvo ECC, please see Ref. 11 by Peter Chudi and Anders Malmquist. In addition, a manufacturer's release for the ECC is available from Olle Boëthius, Environmental Competence Center, Volvo Car Corporation, S-405 08 Göteborg, Sweden.

# The Rosen Motors Turbine-Flywheel Power Train for Automobiles

Another attractive drive system for a hybrid electric automobile has been developed by Rosen Motors in Woodland Hills, California.[12] Founded in 1993 by brothers Benjamin M. Rosen and Harold A. Rosen, they announced on 19 January 1997 on the World Wide Web:

> ...the first successful road test of its experimental hybrid-electric power train for automobiles. This historic road test makes Rosen Motors the only company to have demonstrated the viability of using a flywheel-turbine combination to power an electric automobile.

> "The Rosen TurboFlywheel (RTF) power train has the potential to simultaneously provide unusually long range, nearly zero emissions, and sports car-like acceleration, a combination of characteristics that neither the internal combustion engine nor a battery-powered electric drive train can offer. 'To us, this demonstration is akin to the first flight of a jet

aircraft: proof that our technology works and the first major step to building a superior mode of transportation,' said Dr. Harold Rosen, president and chief executive officer of Rosen Motors. 'This is a milestone in automotive history,' said Benjamin M. Rosen, chairman of Rosen Motors. 'Our hybrid-electric power train, in contrast to some other hybrid-electric concepts under investigation by automobile manufacturers, is not simply a 'range extender' for a battery-powered vehicle. Instead, it is an innovative system that will also sharply cut fossil fuel consumption and measurably improve the performance of the twenty-first century automobile.' The dramatic road test was performed on Sunday January 5 at the Willow Springs racetrack in the Mojave Desert near Edwards Air Force Base. For the initial proof-of-concept, Rosen Motors tested the RTF power train in a converted 1993 Saturn vehicle.

The power train consists of four major elements: a turbogenerator, a flywheel motor-generator, an electric drive motor, and an electronic control system. The turbogenerator, which uses unleaded gasoline as its fuel, produces cruising power for the vehicle. The flywheel, via its motor-generator, supplies surge power for acceleration and recovers power from regenerative braking for later reuse. The electronic controller provides the 'brains' and the power switching for the system. Development efforts are continuing at Rosen Motors to improve the design and characteristics of the RTF power train. Road testing of a more advanced design of the RTF power train, to be housed in a luxury sports sedan, is expected to begin in late 1997.

Future Rosen-powered cars, depending on their weight and aerodynamic characteristics, are expected to obtain more than double the gasoline mileage of today's cars and to provide vehicle acceleration from zero to 60 mph (97 km/h) in six to seven seconds. In addition, future versions of Rosen-powered automobiles are projected to produce tailpipe emissions that are no greater than those caused by the emissions due to recharging battery-powered vehicles from fossil-fueled power plants.

Benjamin M. Rosen and his brother, Dr. Harold A. Rosen, started Rosen Motors after each had many years of success in technology venture capital and communications engineering, respectively. Ben Rosen, chairman of Compaq Computer and co-founder of Sevin Rosen Funds, a venture capital firm, was listed by *Computerworld* magazine in 1992 as one of 25 people in the computer industry who 'changed the world.'

Harold Rosen, a pioneer in communications technology, is recognized as the father of the geostationary communications satellite, the world's first practical commercial communications satellite. He has won numerous prestigious engineering awards in many countries, including the Draper Prize in 1995, the world's top award in engineering. Ben and Harold Rosen are also founding directors of Capstone Turbine Corporation, which supplies the gas-turbine generator, one of the key elements of the Rosen Motors power train. Capstone Turbine is a developer of small gas turbine-driven electric generators (turbogenerators) for both vehicular and stationary power. Privately funded Rosen Motors develops hybrid electric power trains for automobiles and flywheel systems for stationary power applications. Headquartered in Woodland Hills, California, Rosen Motors employs 60 people.

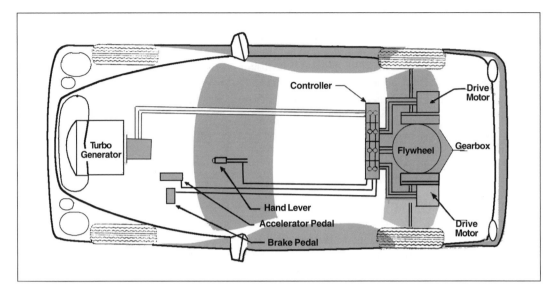

*Figure 7.5. The Rosen drive system. (Rosen Motors)*

Figure 7.5 shows a drawing of this drive system which links turbine and flywheel technologies.

The next chapter treats the Stirling engine, currently being considered for automotive use.

## Note 1—The Whittle Jet Engine

The tremendously powerful jet engines now made in America and other industrial nations for aircraft owe much to Sir Frank Whittle of England. Whittle was born in Coventry on 1 June 1907. He graduated from Leamington College in 1923 to become an apprentice in the Royal Air Force College, then later a cadet, pilot officer, and in 1930 a flying instructor. In 1930, he applied for his first patent for a gas turbine. After a stint at Cambridge University and serving as a test pilot, Whittle joined Power Jets, Ltd. In May 1941, in the midst of World War II, the first flight of an aircraft with a Whittle jet engine was made. Air Commodore Whittle has received many honors.

## Note 2—Chrysler Gas Turbine Cars

In 1954, Chrysler launched a $100 million effort plus $20 million from the federal government on gas turbine-powered, Ghia-built, Chrysler drive train assembled cars.[13] From 1963 to 1966, approximately 203 American families in 133 cities accumulated more than one million miles on these vehicles. These cars each bore a 130-hp engine. The tachometer of each read 22,500 rpm at idle on a 60,000-rpm scale. While driving at 70 mph, "the engine's whine was virtually inaudible. The program was considered an overwhelming success," with the cars largely vibration free, with low maintenance, and in quiet operation. Why did Chrysler give up? Fuel consumption and exhaust emission were two problems, as well as the projected cost of manufacture. The end of this program was caused by hard times at Chrysler in 1981.

# References

1.  Shepherd, D.G., *Introduction to the Gas Turbine*, Van Nostrand Reinhold, New York, 1960.
2.  Murphy, Gordon J., "Appendix B: Three-Phase Induction Motors and Controls for Electric Cars," in Wakefield, Ernest H., *History of the Electric Automobile: Battery-Only Powered Cars,* Society of Automotive Engineers, Warrendale, PA, 1994, pp. 457–470.
3.  Gilbert, R.R., *et al.*, "Flywheel Drive Systems Study," Lockheed Missiles and Space Co., Report LMSC-D246393 to the U.S. Environmental Protection Agency, July 1972.
4.  Dugger, G.L., *et al.*, "Heat-Engine/Mechanical Energy Storage Hybrid Propulsion Systems for Vehicles," Report No. APTD-1344 to U.S. Environmental Protection Agency, Office of Air Programs, Johns Hopkins University, Applied Physics Laboratory, March 1972.
5.  Wakefield, Ernest H., *History of the Electric Automobile: Battery-Only-Powered Cars,* Society of Automotive Engineers, Warrendale, PA, 1994, p. 155.
6.  Suris, Oscar, "Chrysler Corporation Plans to Show a Model of the 'Flywheel' Auto," *The Wall Street Journal*, 13 January 1994, p. 10.
7.  Wald, Matthew L., "Chrysler's Electric Race Car Has Turbine and Flywheel," *The New York Times,* 6 January 1994, p. C16.
8.  Stephenson, R. Rhoads *et al.*, "Should We Have a New Engine?" Technical Report, JPL SP 43-17, Vol. II, Jet Propulsion Laboratory, California Institute of Technology, Pasadena, CA, August 1975.
9.  *Popular Science,* June 1994, p. 101.
10. Valenti, Michael, "Hybrid Car Promises High Performance and Low Emissions," *Mechanical Engineering,* July 1994. Copyright permission belongs to *Mechanical Engineering,* a publication of the American Society of Mechanical Engineers.
11. Chudi, Peter, and Malmquist, Anders, "A Hybrid Drive for the Car of the Future," ABB Corporate Research, P.O. Box 1785, S-11197, Stockholm, Sweden.
12. Castleman, Deborah, Rosen Motors, Woodland Hills, CA, http://www.rosenmotors.com.
13. Sherman, Don, "Retrospect—'63 Chrysler Turbine: Pondering a Pistonless Future," *Motor Trend,* March 1992, pp. 109–110.

# CHAPTER 8

# The Origin, Operation, and Applications of the Stirling Engine

In the course of human events, relatively few motive power sources have been invented. To these prime movers such as the windmill, the waterwheel, the turbine, the steam engine, the internal combustion engine, the battery, the solar cell, the fuel cell, and the nuclear reactor, the hot air engine was added in 1816 to this rarefied list by the Presbyterian minister, the Reverend Dr. Robert Stirling. In honor of its inventor, this prime mover has been named the Stirling engine. Although conceived shortly after the cessation of the Napoleonic Wars, the Stirling engine is now being investigated for use in stand-alone and multi-powered electric vehicles because of its low emissions, its relative silence of operation, and its high efficiency. Below is a brief biography of Reverend Stirling, the historical development period of the Stirling engine, an explanation of its operation, some current applications, and its introduction to the automobile.

## Robert Stirling

Because of Robert Stirling's unique contribution, his background should be noted. Quoting from Graham Walker[1]:

155

Robert Stirling was born in Cloag (Scotland) on October 25, 1790. (See Figure 8.1) He studied at the University of Glasgow, as is mentioned in the Fastie (a type of calendar) but also at Edinburgh University (1805ñ1806 and 1808). In 1805, he took classes in Latin and Greek, and in 1806...advanced Latin and Greek, logic and mathematics, metaphysics, and rhetoric... Robert Stirling was 15 years old when he went to Edinburgh... (he) was licensed to preach by the Presbytery of Dunbarton on 4 July 1815... (and) was presented to the second charge at Kilmarnock, Ayrshire, by the Commissioner of the Duke and Duchess of Portland, and (there) was ordained to the ministry on 19 September 1816. He immediately became second or junior minister of Laigh Kirk, Kilmarnock, and two months later (at 26 years of age) he applied for his first patent for the Stirling air engine and the heat regeneratoróor *economizer*, as he himself called it. The patent assigned to him was No. 4081 of 1816. On 10 July 1819, he married Jane, the eldest daughter of William Rankine, a wine merchant at Galston, and five years later he had been ëtranslatedí to Galston as minister of the church there. In 1827, and again in 1840, he and his brother James repatented the air engine.

*Figure 8.1. Reverend Dr. Robert Stirling (1790–1878). (Photograph of a painting of Robert Stirling in Galston Parish Church, Ayrshire, Scotland. N.V. Philips, Eindhoven, the Netherlands)*

Early in 1840, the University of St. Andrew's, the oldest of Scottish universities, conferred upon him the honorary degree of Doctor of Divinity, probably in recognition of his scholarly attainments rather than for his additional scientific achievements, as is stated in some places. Without doubt, this (recognition) was a glorious day for the minister of Galston, who nonetheless remained in the village of his ministry, gaining widespread respect by the quality of his life and work—a quality shown by his care for the victims of the 1848–1849 cholera epidemic. Ill health at last forced him from the pulpit in 1876, and he died in 1878.

Stirling's tombstone reads:

Rev. Dr. Robert Stirling (1790–1878).
Minister, Church of Scotland, and Inventor Extraordinaire.

Written in 1878, shortly before his death, the following was Stirling's final thought on the prime mover he created:

...These imperfections (of the Stirling engine) have been in a great measure removed by time and especially by the genius of the distinguished (Sir Henry) Bessemer (inventor of the industrial process for making steel from cast iron). If Bessemer iron or steel had been known thirty-five or forty years ago, there is scarce a doubt that the air engine would have been a great success... It remains for some skilled and ambitious mechanic in a future age to repeat it under more favorable circumstances and with complete success...

# Subsequent Development of the Stirling Engine

As if to fulfill Dr. Stirling's writings, Captain John Ericsson (1803–1889), the Swedish-American inventor, revived the Stirling engine in 1853. Not choosing to call it a hot air engine, he designated it a *Caloric engine*. By 1837, Ericsson had designed and patented the propeller for vessel use. Because he was unable to sell propellers in England, a nation whose steamboats at that time were committed to Nicholas J. Roosevelt's 1798 American-patented paddle wheel, Ericsson emigrated to America and continued in marine engineering. In 1853, Ericsson placed a Caloric engine in the 303-ft (92-meter) brig-rigged *Ericsson*.[2] The vessel had four 14-ft (4.3-meter) diameter cylinders which developed 300 hp. This power plant drove the vessel at 11 knots with 32-ft (9.75-meter) diameter paddle wheels rotating at 12 to 13 turns per minute.

Dr. Richard S. Hartenberg[a] continues:

In the spring of 1854, the *Ericsson* returned to New York from Washington, DC, for a refit. She was made ready and took a trip down New York Bay on 15 March 1854, with a second jaunt on April 27... Near Sandy Hook, the ship was struck by a terrific tornado on the port side that careened the hull to put her lower starboard ports, that were open, completely

---

[a] Dr. Richard S. Hartenberg is Professor Emeritus of Mechanical Engineering at Northwestern University, Evanston, Illinois, and is a Registered Engineer. He wrote the paragraph about the *Ericsson* in 1994.

under water. The ship sank, and was raised with the decision to remove the Caloric engines and convert the vessel to steam. Although economical on fuel, the four engines occupied too much revenue space with the four 14-ft (4.3-meter) cylinders.

In 1862, Ericsson subsequently built in 100 days the ironclad *USS Monitor*. This steamship appeared in Hampton Roads, Virginia, the evening after the steam-powered, Confederate armored ram *Virginia* (*Merrimac*) had destroyed three of the largest Union ships, and prepared the next day (9 March 1862) to sink the remaining Union war vessels. The *Merrimac* was met in battle by the *Monitor*. The resulting contest between the ironclads—a draw—altered naval warfare worldwide. A British journalist from the *London Times* observed the contest as the more agile *Monitor*, the first ironclad with a revolving turret, repeatedly struck the *Merrimac*. The journalist dispatched to his newspaper, "Alone, the *Monitor* could sink the entire British navy."

# Eighty Years Later

For approximately 80 years, work on the Stirling engine remained at a low ebb. Meanwhile, the Dutch physicist Heike Kamerlingh Onnes (1856–1926) developed liquefied helium, an achievement for which the Nobel Prize was granted in 1913. In subsequent operations for liquefying helium, N.V. Philips of Eindhoven, the Netherlands, modernized and built several Stirling engines. With increasing concern for pollution in major cities worldwide, the Stirling engine was applied to the automobile by Philips in conjunction with both Ford and General Motors. Because the Stirling engine was such a small part of Philip's total effort, after several years the pertinent employees, led by Dr. Roelf J. Meijer, bought the rights to the Stirling engine from Philips. Dr. Meijer, while at the N.V. Philips Company in Eindhoven, was one of the seminal group who developed the modern Stirling engine. Interest in this prime mover has recently accelerated because of its ability to efficiently deliver substantial mechanical power, with lower emissions. With others, Dr. Meijer founded General Stirling Inc., which has operations in the United States and the Netherlands. The American branch, Stirling Thermal Motors, Inc. (STM), is located in Ann Arbor, Michigan, near the Detroit automotive area. The Netherlands branch, Stirling Cryogenic and Refrigeration (SCR), serves the cryogenic activity of the company. Dr. Meijer is now a Director of General Stirling, Inc., as well as President of Product Innovation Center, Inc. in Saline, a city near Ann Arbor in Michigan.

Dr. Meijer has kindly written the following essay for inclusion in this book.

### The Operation of a Stirling Engine
### Dr. Roelf J. Meijer

#### Introduction

The Stirling engine, formerly called the hot air engine, is based on the closed cycle. The machine was invented by the Scottish minister Robert Stirling in 1816, long before the internal combustion engine was introduced. (Figure 8.2 shows an early version.) Hot air engines were used in considerable numbers during the nineteenth century, and it was thought for a time this type of prime mover might rival the steam engine as a source of power. However, the Stirling engines then were enormously large—2000 to 4000 lb (900 to 1800 kg) per horsepower. Moreover, their efficiency was low although better than the steam engine. Small wonder the internal combustion engine, with its much greater specific power and higher efficiency, displaced this hot air engine despite a certain charm which the latter possessed.

After the internal combustion engine reached maturity, showing its advantages and disadvantages, intensive research on the Stirling hot air engine was carried out in the Research Laboratories of N.V. Philips in the Netherlands for more than 40 years, with the end result being the 175-hp engine for the *Torino* automobile of the Ford Motor Company. This engine was delivered to Ford in 1976 and is shown in Figure 8.3.

### The Principle of the Stirling Engine

An internal combustion engine provides a surplus of work by virtue of the compression at low temperature of a certain quantity of air, to which atomized fuel is added either before or after its compression, the subsequent heating of the mixture by rapid combustion, and its expansion at high temperature. The Stirling engine is based on the same principle (i.e., the

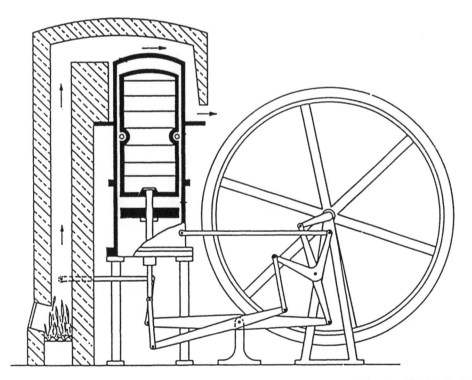

*Figure 8.2. The first Stirling engine according to the patent of 1816. (Roelf J. Meijer)*

*Figure 8.3. The 175-hp double-acting Stirling engine for the Ford Torino (1976), now in the automotive museum in the Netherlands. (Roelf J. Meijer)*

compression at low temperature and expansion at high temperature of a given quantity of gas). However, the heating occurs in an entirely different manner, heat being supplied to the gas from outside, through a wall, due to the nature of the closed cycle.

Of all the Stirling engine configurations known, three can be distinguished: heat exchangers (cooler, regenerator, and heater) and two variable volumes, one on the side of the cooler (cold space) and one on the side of the heater (hot space). The regenerator is placed between the cooler and the heater. (See Figure 8.4) Because of a certain phase difference between these variable volumes, the working gas is shuttled from the hot space, which is at constant high temperature (e.g., 370°F [700°C]), into the cold space, which is at constant low temperature (e.g., ambient temperature). To obtain mechanical energy from this process, the working gas must be compressed when most of it is in the cold space and must be allowed to expand when most of it is in the hot space. To prevent heat loss during this shuttling process, the regenerator, which is a space filled with porous material such as layers of very fine gauze, captures the heat of the gas as it flows from the hot space to the cold space and returns this heat to the gas on its way back to the hot space.

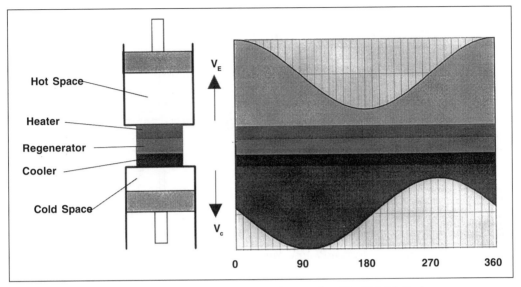

Hot Space

Heater

Regenerator

Cooler

Cold Space

$V_E$

$V_c$

0    90    180    270    360

*Figure 8.4. Illustration of the Stirling cycle. (Roelf J. Meijer)*

**Development of the Stirling Engine**

To make the Stirling engine compact, reliable, and highly efficient, many configurations have existed. However, three have received the most emphasis:

**1)The Displacer Type.** This was the original type developed by Stirling, but with a rhombic drive, as shown in Figure 8.5. The single-cylinder rhombic drive, which is 100% balanced, is a good tool for Stirling investigations. This configuration, together with the introduction of hydrogen and helium gas and the development of a compact heat exchanger, made possible an engine with specific power and efficiency equal to the diesel engine. This development occurred between 1955 and 1970.

**2)The Double-Acting System.** A modern Stirling engine is double-acting and can be a three-, four-, five-, or six-cylinder engine. A four-cylinder type is shown schematically in Figure 8.6. The compression of the gas takes place under the pistons (the cold space) and its expansion on top of the pistons (the hot space), and it is accomplished by the movement of the pistons, which have a 90° phase difference. After the expansion of the gas in the hot space, the gas is shuttled, via the heater regenerator and cooler, into the cold space. There the gas is compressed and forced back to the hot space. Completion of one cycle of the engine results in one revolution and an excess of work.

An example of such a cycle is the Ford *Torino* engine shown in Figure 8.3. The swash plate drive of this engine makes it compact, resulting in a specific power approaching the conventional gasoline engine. The Ford *Torino* engine has shown that the swash plate drive is very suitable for a Stirling engine.

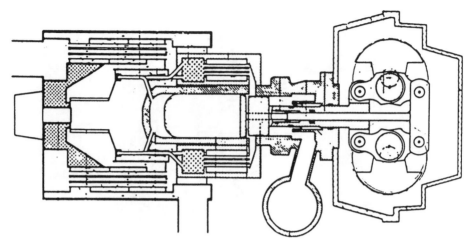

*Figure 8.5. Cross section of a Philips' Stirling rhombic drive engine. (Roelf J. Meijer)*

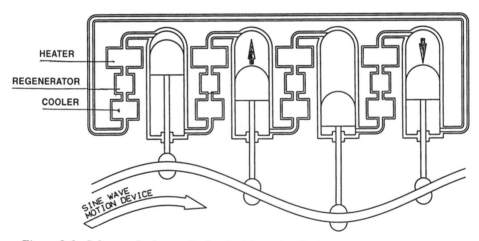

*Figure 8.6. Scheme of a four-cylinder double-acting Stirling engine. (Roelf J. Meijer)*

The closed cycle gives the Stirling engine special properties, such as relatively silent operation, very low exhaust pollution, and multi-fuel capability. However, the disadvantage of the closed cycle is that the system must be hermetically sealed if gas other than air is used as the working fluid. Also, the power control caused by pumping gas in and out of the cycle was cumbersome. A breakthrough came in 1980 when STM introduced the variable swash plate drive by which the power control was accomplished quickly by mechanically changing the swash plate angle and thus the swept volume of the piston. This modification allowed the use of a pressure hull around the compact drive by which only the rotating shaft must be sealed by a commercially available seal. See Figure 8.7.

Today, STM is involved with one of the Big Three automotive companies in developing a lightweight, highly efficient engine for a hybrid electric vehicle.

**3) The Free-Piston Engine.** This engine, introduced in the 1960s by Professor William Beal, is completely closed but in general serves a much smaller power output. A generator set, a linear alternator, is enclosed in the same pressure vessel. See Figure 8.8.

**Bibliography**

1. Rinia, H., and du Pré, F.K., "Air Engines," *Philips Technical Review,* **8**, 1946, pp. 129–136.

2. Meijer, R.J., "The Philips Hot-Gas Engine with Rhombic Drive Mechanism," *Philips Technical Review*, **9**, 1958, pp. 245–262.

3. Kitzner, K.W., "Automotive Stirling Engine Development," NASA Report No. CR-159836, March 1980.

4. Meijer, R.J., and Ziph, B., "Variable Displacement Stirling Automobile Power Trains," Proceedings of the Fifth International Automotive Propulsion Systems Symposium, Dearborn, MI, April 1980.

*Figure 8.7. The STM4-120 engine with variable swash plate drive and pressure hull. (Roelf J. Meijer)*

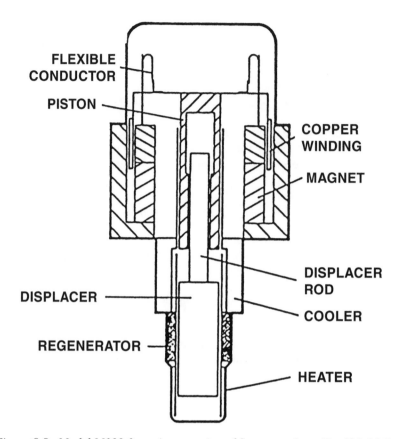

*Figure 8.8. Model M100 free-piston engine of Sunpower Inc. (Roelf J. Meijer)*

5. Meijer, R.J., "The STM4-120RH and the STM4-120 DH Stirling Engines' Performances," Korea Society of Automotive Engineers, Sixth International Pacific Conference on Automotive Engineering, 1991.

6. Urieli, I., and Berschowitz, D.M., *Stirling Cycle Engine Analysis*, Hilger Ltd., Bristol, UK.

In addition to Dr. Meijer's essay, further information on the Stirling engine may be found in Ref. 1 and Refs. 3 through 6 of this chapter.

## Sterling Thermal Motors' Solar Power Conversion

While the Presbyterian minister's Stirling engine is old in principle, it is being seriously considered today by at least one of the Big Three American automotive manufacturers as an Auxiliary Power Unit (APU) for multi-powered electric cars. An actual STM industrial model is described in the following literature from Stirling Thermal Motors Inc.[7]:

The STM-120 Stirling engine is the result of 15 years of Stirling engine development at Stirling Thermal Motors, Inc.. (STM) in Ann Arbor, Michigan. The four-cylinder, double-acting STM4-120 Stirling engine is optimized to produce 25 kW at 1800 rpm and is capable of producing 52 kW at 4500 rpm. This engine has demonstrated performance and fuel efficiency equivalent to a diesel engine, with ultra-low emissions and low noise and vibration level. The engine's environmentally friendly qualities, multi-fuel capabilities, and potential for long service-free life have generated interest in manufacturing and marketing this engine for commercial applications. (The system is shown in Figure 8.9.)

The STM4-120 engine can be used for a broad variety of applications by simply changing the receiver/heater head to use various types of heat sources such as combustion of liquid or gaseous fuel. This results in a mass-produced basic engine with lower manufacturing costs. The manufacturing cost will be equal to a similar-sized diesel engine.

**Solar thermal power applications** are an immediate opportunity for use of the STM4-120 Stirling engine. An STM Solar Power Conversion System (PCS) has been undergoing on-sun testing at Sandia National Laboratories since February 1993 (Figure 8.10). In November 1993, Sandia awarded a contract to Science Applications International Corporation (SAIC) to develop for commercial use a 25-kW Utility Scale Solar Thermal Power System.

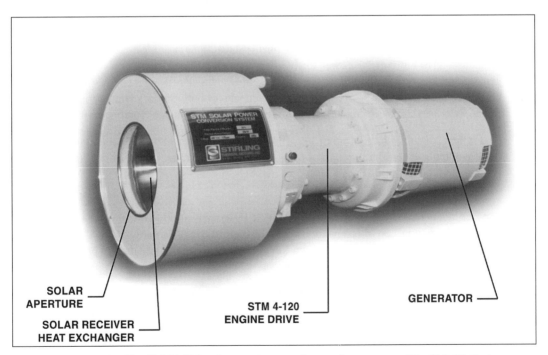

SOLAR APERTURE

SOLAR RECEIVER HEAT EXCHANGER

STM 4-120 ENGINE DRIVE

GENERATOR

*Figure 8.9. The STM4-120 solar receiver, engine, and generator. (Roelf J. Meijer)*

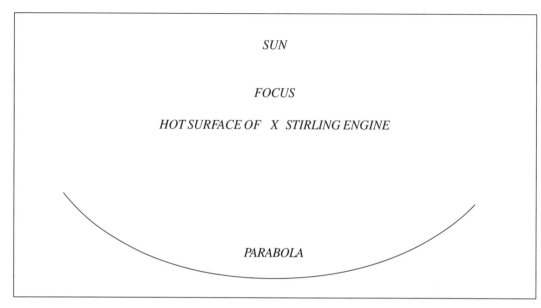

*Figure 8.10.  Obtaining electricity from the reflected sun's rays using the STM4-120.*

The five-year, three-phase program has a potential value of $35 million.  Half the program is funded by DOE (the U.S. Department of Energy) and half by the industry participants. The project has a goal of commercializing the technology by the late 1990s.  STM is subcontractor to SAIC and is responsible for the development and supply of the high-efficiency STM4-120 Solar Power Conversion System.  The system has a potential to generate electricity at about three to six cents per kilowatt hour, making it competitive with (many) conventional power sources without releasing air pollution or greenhouse gases. The system also can be operated with natural gas at night or during cloudy periods.

Under the program, a total of 56 dish/Stirling power systems will be manufactured and demonstrated at electric utility sites throughout the United States.  The system then will be marketed to U.S. utilities as a pollution-free distributed power system and to foreign countries for utility and rural electrification.

The DOE/General Motors Hybrid Electric Vehicle Program is a good example of the multi-application potential.  STM is developing a multi-fuel Auxiliary Power Unit (APU) under subcontract to GM.  STM is currently working on the design and fabrication of prototype APUs to be installed and tested in a GM hybrid vehicle.

Superior performance and manufacturability of the STM4-120 are a direct result of the unique variable displacement power control development at STM.  While many Stirling engines are efficient at a specific operating point, variable displacement variation is inherently more efficient than variable pressure.  Changing engine power by varying piston stroke instead of the pressure of the working gas provides several advantages:

- Exceptional part-load performance: nearly constant efficiency over 75% of the engine power range and fast engine response.

- Improved reliability, reduced cost, and elimination of working gas compressor, storage tank, and external valves.

In addition to the experimentation with the Stirling engine cited above, General Motors was active in Detroit. See Table 8.1 for technical specifications of the STM4-120.

## TABLE 8.1
## TECHNICAL SPECIFICATIONS OF THE STM4-120

| | | | |
|---|---|---|---|
| **Receiver** | | **Cooling System** | |
| Type: | Directly Illuminated Tubes | Type: | Water/Air Radiators |
| Aperture Diameter: | 220 mm (8.66 in.) | Capacity: | 40 kW$_{th}$ (136,500 Btu/hr) |
| Incident Flux (Peak): | 75 W/cm$^2$ (24 Btu/hrft$^2$) | Coolant: | 90–10% Water/Glycol |
| Max. Tube Temperature: | 800°C(1470°F) | Water Pump: | 1.3 l/s (21 gpm) 200 W |
| | | Fan: | 2.8 m$^3$/sec (6000 cfm) |
| **Stirling Engine** | | **PCS Dimensions** | |
| Type: | STM4-120 | Height: | 0.86 m (34 in.) |
| Working Gas: | Helium or Hydrogen | Width: | 0.86 m (34 in.) |
| Power/Speed: | 26 kW (35 hp)/1800 rpm | Length: | 1.98 m (78 in.) |
| | 52 kW (68 hp)/4500 rpm | | |
| | | PCS Weight: | 725 kg (1596 lb) |
| Thermal Efficiency: | 43% | | |
| Power Control: | Variable Displacement | **Generator:** | |
| Number of Cylinders: | 4 | Type: | Induction |
| Swept Volume: | 120 cc (7.3 in$^3$)/Cylinder | Power Output: | 20 kWe |
| Working Gas Pressure: | 12 MPa (1740 psi) | Voltage: | 480VAC/Three-Phase |
| Working Gas Temperature: | 720°C (1325°F) | Frequency: | 60 Hz |

# General Motors *Stir-Lec I*

For the *Stir-Lec I* development, General Motors chose P.D. Agarwal to head the group.[8] Agrawal had led the previous team that produced the remarkable *Electrovair* and the *Electrovan,* early three-phase a-c drive vehicles described in Chapter 19 of *History of the Electric Automobile: Battery-Only Powered Cars.*[9] Again, R. Rhoads Stephenson *et al.* continue[10]:

> The General Motors *Stir-Lec I* is a series battery hybrid which uses a previously developed 8-hp Stirling engine weighing 450 lb (204 kg) with approximately 500 lb (227 kg) of starting-lighting-ignition (SLI) lead-acid batteries. The total power train weight of 1700 lb (771 kg), including batteries, results in a vehicle (with a) curb weight of 3200 lb (1452 kg)

167

when mounted in an Opel *Kadett* body. A 20-hp lightweight liquid-cooled induction motor drives the differential directly without the use of a transmission. A three-phase controller regulates power to the motor from the batteries... Engine power is delivered to the battery through a 25-hp three-phase alternator-rectifier system, which is greatly over-rated for the 8-hp Stirling engine output. (The road performance characteristics are shown in Table 8.2.)

The system produces excellent performance with respect to CO and HC emissions. General Motors feels that acceptable $NO_x$ performance can be achieved by engine modifications.

### TABLE 8.2
### *STIR-LEC I* PERFORMANCE

|  | Battery Only | Engine and Battery |
| --- | --- | --- |
| Level road fuel economy at 30 mph (50 km/h) | — | 30–40 mpg (12–18 km/liter) |
| Level road driving cycle range | 15–30 miles (25–50 km) | — |
| Top speed | 30 mph (50 km/h) | 55 mph (90 km/h) |
| Range at top speed 55 mph (90 km/h) | — | 30–40 miles (50–70 km) |
| 0–30 (0–50) acceleration time | 6 seconds | 6 seconds |

The phantom view and block diagram of the *Stir-Lec I* are shown in Figure 8.11.

For further information on the Stirling engine, many additional modern references are available, such as those cited at the conclusion of Dr. Meijer's essay within this chapter.

# Stephenson's Conclusions on Hybrid Electric Vehicles

As described below, R. Rhoads Stephenson *et al.*[10] are not overly sanguine in substituting hybrid drive systems for the conventional drive system.[b] They write:

> Hybrids do not offer greater adaptability to changing fuel availability than do conventional vehicles. The same engine types are used in each, and similar considerations of cost, weight, volume, and BSFC[c] govern the selection of heat engines for each application.

> While the exhaust emissions of hybrids are somewhat lower than those of a conventional vehicle for equivalent road performance, the magnitude of the reduction is not sufficient to meet the 1977 federal emission standards by hybrid operation alone, and no substantial

---

[b] Dr. Victor Wouk seriously challenges the conclusions of Stephenson's report in his letter to me, dated 2 April 1994.

[c] BSFC = Brake Specific Fuel Consumption. Fuel consumption under constant load. See Ref. 11.

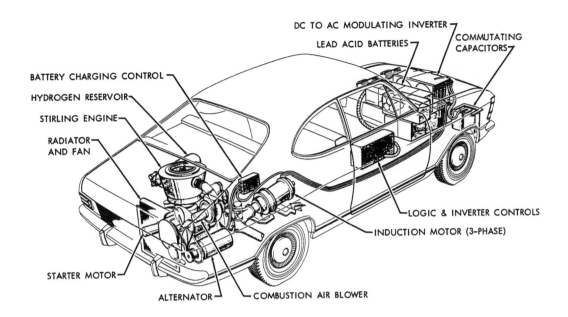

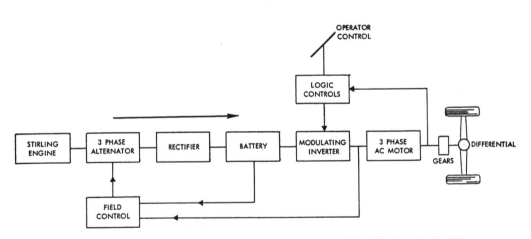

*Figure 8.11. Phantom view and block diagram of the* Stir-Lec I. *(Ref. 8)*

decrease in the number, weight, cost, size, or performance of engine modifications or exhaust after treatment devices can be expected by hybrid operation.

The road performance of hybrids can be made comparable (but not identical) to that of conventional vehicles in all important respects. Greater onboard total peak power capability is required to overcome the greater weight of the hybrid, but less of the total power capacity must be supplied by the heat engine. Thus, smaller, harder-working heat engines suffice for normal passenger car usage. Since reduced heat engine power capacity is essential to achieving the fuel economy advantages of hybrid operation, continuous high power is not available for nontypical driving demands such as trailer-towing or sustained high speed (in excess of 80 mph [129 km/h]) or hill climbing operation.

Based on the calculations in this chapter, the most optimistic projection of fuel economy improvement of hybrids of conventional cars is about 25% under urban driving conditions, assuming that each type of vehicle uses an Otto cycle engine and that the road performances of both are comparable. Under high-speed highway driving conditions, the fuel economy advantage of hybrids nearly disappears, and improvements of 10% may be experienced at best. Of the two basic hybrid configurations, the parallel system offers the best performance in all respects, including fuel economy and life-cycle costs.

The development of advanced engines such as the Stirling which have improved part load fuel economy over the Otto engine will not greatly benefit hybrid operation but will improve the fuel economy of conventional vehicles. Thus, the comparative advantages of hybrid operation will further diminish. Finally, the combination of a small engine (like that used in hybrids) with a CVT (continuously variable transmission) in a conventional vehicle configuration will provide generally equivalent road performance to that of a hybrid (somewhat poorer instantaneous acceleration and short grade hill climbing power, but better sustained high speed and long grade hill climbing performance) at improved overall fuel economy. In comparison with a conventional car with the same small size of engine, the CVT vehicle will produce substantially improved road performance with only slightly poorer fuel economy.

A rapid buildup of the battery hybrid (vehicle) population would produce a severe impact on the materials industry. Substantial production rate increases for critical materials such as lead, nickel, or zinc would be required for a buildup in hybrid production to 10 million vehicles/year. Flywheel hybrids, on the other hand, place minimal requirements on critical material supplies.

The initial cost of all types of hybrids will remain higher than that of conventional vehicles of equal performance and accommodations due to their added weight and complexity. The life-cycle cost will also remain substantially greater until such time as fuel cost savings which result from improved fuel economy exceed the amortized increase in the initial cost plus the additional maintenance cost. Based on the estimates in this chapter, this point will not be reached until the cost of gasoline increases beyond $1.70 per gallon.

That quote was written by R. Rhoads Stephenson *et al.* in 1975. Dr. Victor Wouk, a long-time advocate of hybrid automobiles, would have a different conclusion.

# Notes

A conceptual study on hybrid electric vehicles which should be noted is one that includes a hydraulic pump used essentially as a regenerative brake in electrical terms.[12] The vehicle has as prime mover a Stirling engine/generator and a set of batteries. Figures 8.12 and 8.13 are schematics of the system; the components are shown in Table 8.3. For additional information, refer to the report.[12]

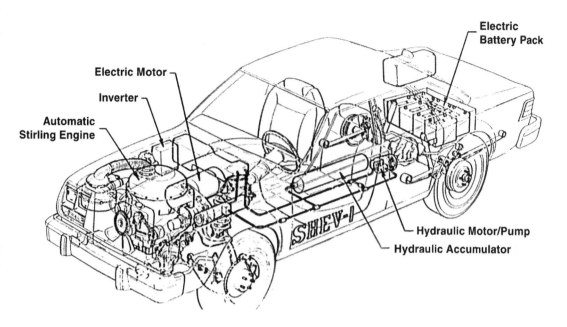

*Figure 8.12. Parallel system with short-term storage. (Ref. 12)*

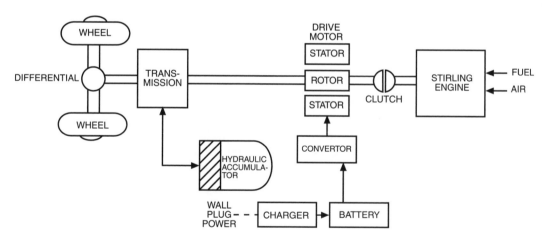

*Figure 8.13. Conceptual layout of advanced propulsion system. (Ref. 12)*

**TABLE 8.3**
**COMPONENTS OF THE MECHANICAL TECHNOLOGY INC.**
**HYBRID ELECTRIC VEHICLE**

| | |
|---|---|
| • Engine: | Kinematic Stirling<br>10.5 kW average operating point<br>(peak efficiency point) |
| • Batteries: | Seven 12-volt lead-acid |
| • Motor: | a-c three-phase, induction<br>12.8 kW<br>8,000 r/m in (rpm)<br>20 cm wide × 38 cm long (8 in. × 15 in.) |
| • Inverter: | Variable frequency controller<br>45 cm × 50 cm × 20 cm (18 in. × 20 in. × 8 in.) |
| • Hydraulic Motor/Pump: | 36.4 kW<br>0.045 m³ (1.6 ft³) volume with accumulator |
| • Microprocessor Control: | 15 cm × 15 cm × 8 cm (6 in. × 6 in. × 3 in.) |

The next chapter introduces the thinking and experimentation that set the stage for the application of photovoltaic cells to solar-powered and solar-assist electric-powered vehicles.

# References

1. Walker, Graham, *Stirling Engine*, Clarendon Press, Oxford, UK, 1980, by permission of Oxford University Press.
2. Ericsson's Caloric Engine. Articles descriptive of the caloric ship Ericsson, and of her trial excursion of January 12, 1853; taken from the daily journals of the city of New York, Gideon & Co., Printers, Washington, 1853, University of Michigan Library. See also: *Scientific American*, **VIII**, 20, 29 January 1853, p. 160.
3. Zarinchang, J., *The Stirling Engine*, Intermediate Technology Group, London, 1972.
4. Ackerman, Robert A., "Free Piston Stirling Engine Development," Final Report (1 January 1986–31 December 1986), Gas Research Institute, Chicago, 1988.
5. Walker, Graham, *Stirling Cycle Machines*, Clarendon Press, UK, 1974.
6. Collie, M.J., *Stirling Engine Design and Feasibility for Automotive Use*, Noyes Data Corp., Park Ridge, NJ, 1979.
7. Johannson, Hennart, President, Stirling Thermal Motors, Inc., Ann Arbor, MI.

8.  Agarwal, P.D., *et al., "Stir-Lec I,* A Stirling-Electric Hybrid Car," SAE Paper No. 690074, Society of Automotive Engineers, Warrendale, PA, 1969.

9.  Wakefield, Ernest H., *History of the Electric Automobile: Battery-Only Powered Cars*, Society of Automotive Engineers, Warrendale, PA, 1994.

10. R. Rhoads Stephenson *et al.,* "Should We Have a New Engine?" Technical Report, JPL SP 43-17, Vol. II, Jet Propulsion Laboratory, California Institute of Technology, Pasadena, CA, August 1975, Chapter 9, p. 3.

11. Ribbens, William B., and Mansour, Norman P., *Understanding Automotive Electronics*, Howard W. Sams & Company, Carmel, IN, 1990, p. 173.

12. Bhate, Suresh, Chen, Hsin, and Dochat, George, "Advanced Propulsion System Concept for Hybrid Vehicles," prepared for Stirling Energy Division, National Aeronautics and Space Administration, Lewis Research Center, Contract DEN 3-92, December 1980.

# CHAPTER 9

# On the Road to an Emissionless Automobile

## Competing with Gasoline-Powered Cars

When will there be an emissionless automobile which will compete with a petroleum-powered motorcar, specifically in range, in ease and cost of refueling, and in widespread and convenient infrastructure? Nobody knows the answer to that simple question, although an enormous amount of time and funds has been expended during the past 30 years. The gasoline-powered automobile has proven a world-beater since approximately 1902 when those most perceptive judged the competing vehicles: gasoline-, steam-, and electric-powered. The gasoline-powered vehicle was superior and has so remained.

However, some optimists think a convergence of events currently is forming which may result in an electric vehicle capable of reasonably large-scale production and capable of being sold in substantial numbers.[a] In short, a small, widening window of opportunity appears to exist for the electric car, a window which may gradually grow larger. Among these converging events, we might list the following:

1. The State of California regulation insisting that 2% of cars sold in California in 1998, and a greater percentage in later years, must be emissionless (roughly 40,000 cars per year in 1998, since modified).

---

[a] According to Jim Matega, *Tribune* Auto Writer, "...76 battery-powered EV-1 cars were leased from Saturn dealers in California and Arizona in the month they were offered in 1996, and 469 consumers are (now) negotiating a lease for the vehicles." *Chicago Tribune,* 26 January 1997, Sec. 12, pp. 1 and 7.

2.  The progress in electric battery development, plus some other compatible means of energy storage as outlined in the preceding chapters.

3.  The availability of lightweight electric motors and solid-state controls, with both high efficiency and great reliability.

4.  The emergence of a lightweight chassis and body with low drag, and low-rolling-resistance tires, all capable of meeting federal safety standards.

5.  The announced intention of entrepreneur Robert C. Stempel, former CEO of General Motors, and others to help make such cars available. This chapter outlines the importance of a personality such as Stempel and the events that have thrust him forward.

I wrote the following to Mr. Howell Raines, editor of the Editorial page of the *New York Times*, on 31 January 1994:

> The January 25 issue of the *New York Times* indicated the former Chairman of General Motors, Robert Stempel, plans to offer electric cars from a new company.[1] To reach this decision, what are some of the problems he must have considered?
>
> 1.  There have been more than 200 electric car companies. Few lived long. The gasoline-powered car offers highly developed and convenient travel.
>
> 2.  One pound of gasoline contains 300 to 400 times more applicable energy than one pound of lead-acid battery. The tank of a gasoline car can be filled in four to five minutes. The charging of a battery requires a substantially longer time.
>
> But a small window for a company to manufacture electric cars does exist. Both the State of California regulations for 1998 and the reluctance of the Big Three to formally launch their electric autos has opened this window.
>
> Favoring Mr. Stempel are: 1) He knows conventional automobiles, and he knows electric cars. Moreover, he has central knowledge of personnel in both industries, and surely he has access to large capital. 2) The increasing reluctance of the Big Three is his trump card. Each may aid his success, even to placing their nameplates on his cars. Each, with its nationwide sales and repair facilities, can then claim electric cars, thus mitigating the onus and increasing psychological pressure to offer electric motorcars, *now.* Further, as the reference (Ref. 2) below cites, Japanese companies have active research projects on electrics. (They also have like problems.)
>
> Inasmuch as electric vehicles will be niche players because of limited range, what type car might Mr. Stempel offer: a sedan or a small van? GM has shown a two-place sedan, *Impact,* with lead-acid batteries; Ford, a small van, *Ecostar,* with sodium-sulfur batteries; and Chrysler, a large van with nickel-iron batteries. All, as well designed as they are, represent compromises. In range, none match the gasoline-powered car.

What about government regulations as to safety? If Mr. Stempel builds a ground-up car—a *compleat vehicle* in the vernacular of the trade—he faces a thicket of regulations. If he chooses a conversion, his path is considerably simplified, but there are more compromises.

What of the drive system? Alternating-current (a-c), or direct-current (d-c)? The world has turned to a-c drives. And what type of motor? He has (several) choices. What should be his basic battery voltage? And how should the batteries (a heavyweight) be placed? These questions are few of many. Finally, Mr. Stempel may be the best man in America to introduce a truly viable electric car. If he fails in offering an emissionless viable car in quantity, California may need to modify its controversial edict. (See Ref. 3)

# Opening the Window for Emissionless Cars

Strangely enough, we can reason, opening a window of opportunity for truly viable emissionless electric cars begins in Australia. As you continue reading this chapter, you will understand the connection between Australia and the emissionless car.

To better appreciate the importance of Australia in encouraging the emissionless automobile, a little history is necessary. In this story Mr. Dennis Bartell, a lifelong citizen of this country 'down under,' had a historical interest in traversing this island continent. Bartell knew that on 24 July 1862, the time of the American Civil War, a Scot,

> John McDouall Stuart, after five false starts over thirteen lonely years, with nine associates, stood looking north onto the Arafura Sea. Stuart, on his grey mare Polly, raised a toast to their nine-month, 2000-mile trek from Adelaide in South Australia. In what is now Darwin, Australia, the men blazed a tree, buried their papers, then walked and rode back to Adelaide.

Figure 9.1 illustrates Stuart's 1862 trek.

As Bill Tuckey continues,[4]

> His men carried him, ulcerous, crippled with scurvy and going blind, in a hammock slung between two horses for the last three months. Back home, he was given 2000 (English) pounds and 2000 square miles of grazing country by a grateful government, a Royal Geographic gold medal, and a presentation watch. Two years later, he died in England, blind. His gravestone at Kensal Green bears one brief claim: 'First Australian explorer to cross the Australian continent from south to north.' He was just 50 years old.
>
> The road that followed him was little more than a dirt track, marking the route of the parties that 10 years later, after one year and 11 months, finally joined the Overland Telegraph line. Thus on 15 November 1872, with an undersea cable from Darwin to what was then Java, Australia was finally joined to England through what the Aborigines called 'The Singing String' of 36,000 telegraph poles and nine stations at a cost of 479,174 pounds, 18 shillings, and three pence. The staggering task was supervised by a young man called Charles

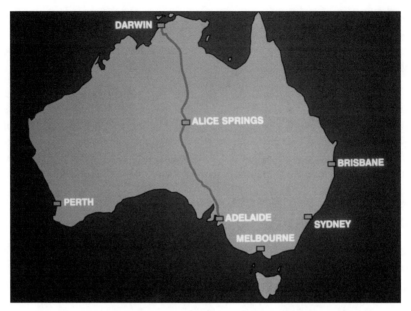

*Figure 9.1. Stuart's trek across Australia in 1862.* (The Advertiser, *Adelaide, Australia*)

Todd, Superintendent of Telegraphs in South Australia, married to Alice Gilliam Bell, whom he first met in the university city of Cambridge when he was 22 and she was 12. She had said to him then, "I will marry you, Mr. Todd." He later named Alice Springs, one of the first telegraph stations and now a city of 30,000 halfway between Darwin and Adelaide, after her.

For South Australia's "Jubilee 150," Bartell determined to re-enact the trek of Stuart's return by walking from Darwin, lying by the Timor Sea, more than 1800 miles (2900 km) to Adelaide in the south, bordering the Indian Ocean. As you know and as Captain James Cook surmised as early as 1770 when he explored the Australian eastern coast, Australia is a vast land with an area almost the same as the 48 contiguous American states, and it has a desert-like interior. Australians call this desert region the 'Outback.' Bartell's trek, re-enacting Stuart's return, required four months following the then building Stuart Highway connecting Darwin with Adelaide by way of Alice Springs, a city lying almost in the geographical center of the continent.

Exposed to a burning sun nearly every day, Bartell had ample time to wonder how the nation of Australia could apply the daily focused solar energy. Aware of the recent development in photovoltaic cells, metallic surfaces that convert the sun's energy into useful electricity, Bartell's mind conjured an electrically powered automobile whose continuous power source would be the sun.

Concluding his trek to Adelaide in the south, Bartell convinced his associates to raise the necessary funds to have Terry Trevor design and the Pecan Engineering Company build an electric motorcar whose electric battery would be continually charged by the output from the photocells. This early solar-powered car was an original design. That is, the Pecan Company installed four lead-acid batteries in a suitable lightweight chassis. The battery fed power to the electric motor which, with the photovoltaic cells, would provide electricity to a d-c motor. In turn, this source of torque, through a transmission, would power the drive wheels of the car. Bartell's batteries could be charged, at a rate of 810 watts with zenith sun, with 5 kWh of electricity, an amount in energy equivalent to the energy stored in approximately one-half gallon of gasoline. However, during average daylight hours, the photocells would recharge the battery at the rate of approximately one-half or more kilowatts. Bartell and his car are shown in Chapter 13.

With only the average sun as a source of power, Bartell could travel approximately 70 miles (113 km) per day. Starting his solar-powered car in Darwin on 11 November 1986, he attended a reception in Adelaide on 5 December, to great acclaim. Bartell had made the first solar-powered car crossing of Australia, north to south.

## The Influence of Hans Tholstrup

Observing this feat, Danish-born Hans Tholstrup made a decision. Tholstrup was a consummate adventurer who had first piloted an outboard-engine-equipped boat a distance of approximately 6000 miles (9656 km) around the circumference of Australia and flew a light plane alone around the world. He, with Larry and Garry Perkins, would form a group which would be the first to pilot a solar-powered automobile across the continent from Perth on the west coast of Australia to Sydney, a major city on the east coast. This vehicle, shown in Chapter 12, made the journey, averaging approximately 20 mph (32 km/h).[5] Having completed this first, Tholstrup, knowing of the solar car races in Switzerland (the Swiss *Tour de Sol* ), asked, why not offer in Australia a true World Solar Challenge race that is open to competitors from all over the world, thus providing the ultimate challenge of racing from Darwin to Adelaide? Figure 9.2 shows Tholstrup.

With great effort, Tholstrup achieved financial backing and a perennial $20,000 cup to provide the first such race to be held in the early weeks of November 1987, the month in the southern hemisphere when daylight is longest. The race would be run every third year. In soliciting participants from all over the world, Tholstrup sent a 14-page announcement and specifications for the race to Mr. Roger Smith, then Chairman of General Motors. Surprising to many, Smith sent the announcement to the Hughes Electronic Group of GM, and his company accepted the challenge. Orders were issued to build the best solar-powered vehicle that experts from all over the world could deliver. On completion, this vehicle was named *Sunraycer.*

*Figure 9.2. Hans Tholstrup conceived and executed the 1987 World Solar Challenge race. Here he is shown with his Honda solar motor bike in 1990. (*The Advertiser, *Adelaide, Australia; photo by Grant Nowell)*

One of the chief designers of *Sunraycer* was Dr. Paul B. MacCready, a California Institute of Technology professor who had already designed and supervised the flying of the first human-powered aircraft and thus claimed a prize for the first aircraft to perform this feat.[6] In addition, MacCready had designed the first solar-powered aircraft to fly the English Channel.[6] Further, MacCready owned a small, highly qualified company, AeroVironment Inc., capable of fabricating unusual products. This company was to be the "skunk works" for General Motors. Never publicized was the amount General Motors spent on building two *Sunraycers,* but the rumor was that "it was in the millions of dollars." The *Sunraycer* was entered and won the first World Solar Challenge race at an average speed of slightly greater than 42 mph

(68 km/h), winning over the second place finisher by approximately two days. Details of this race may be found in Chapter 13 and in Ref. 6. *Sunraycer,* a superb endeavor shown in Chapter 12, now rests in the Smithsonian Institute in Washington, DC.

# Robert Stempel Becomes Chairman and CEO of General Motors

Meanwhile, on the economic front, General Motors was continually and seriously losing market-share of the American auto market. As a result, Roger Smith retired as GM Chairman, and Robert Stempel was promoted. Thus, Stempel was Chairman and CEO of General Motors at the time of the World Solar Challenge Race of 1987. As a result, he participated fully in this event.

Possibly unknown at the time by Stempel, this Australian race would prove one of the more important indices in his life, for little did he realize how short would be his tenure as CEO of General Motors. With the company hemorrhaging in red ink, the GM Board of Directors replaced Stempel, but only after he had created and carried out the General Motors *University Sunrayce USA 1990,* a race of college-built solar-powered cars. This race ran from Lake Buena Vista in Florida to the General Motors Research Center in Warren, Michigan. It was conceived to promote science education in North American colleges and universities. In execution, it provided funding and leadership for solar car racing competition among 32 colleges and universities.

Plans were submitted to a board of judges composed of personnel from: 1) General Motors, 2) MacCready's company, AeroVironment, 3) the U.S. Department of Energy, and 4) the Society of Automotive Engineers. For those 32 schools' plans that were judged best, each school was given a small stipend—so-called seed money—by the U.S. Department of Energy. Since 1990, the balance of funds, very much larger, for each university's solar car, is raised by students. These funds have ranged from several tens of thousands of dollars to nearly a million dollars. The time spent by the students in designing, building, and testing their handwork, in the case of the University of Michigan, has amounted to approximately 25,000 hours.[7]

The *Sunrayce USA* is held every three years, starting in 1990, from a southern location in the United States to a northern city, distant by an amount approximately equal to the distance from Darwin in Australia to Adelaide. This northern hemisphere *Sunrayce USA* race, held in June to match the Australian race in the following November, provides Robert Stempel with first-hand knowledge of this specialized activity, a practice he has maintained although he is no longer associated with General Motors. A lifetime of engineering and management work at GM and extensive experience in electric vehicles of many types are two of the reasons to include Robert Stempel as part of the important converging events opening a window to a possible emissionless, viable electric car for the industrial nations of the world. In 1997 Stempel, shown in Chapter 14, is Chairman of Energy Conversion Devices, Inc. and Ovonic Battery Company, Inc.

The next chapter outlines energy storage elements which, combined with electric batteries, might make viable an emission-free electric automobile.

# Notes

Dennis Bartell's dream of more effectively using the bonanza of solar energy falling on Australia was microanalyzed by a statement made by Dr. Paul B. MacCready in Lecture 1-1 in the Society of Automotive Engineers publication *GM Sunraycer Case History*.[6] Only a small "10% of the energy powering the *Sunraycer* came from the battery, but that 10% was vital for maintaining speed under cloud shadows and while ascending hills."

Using published figures that *Sunraycer* consumes 30 watt-hours per mile while traveling at 50 mph (83 km/h), that its average speed from Darwin to Adelaide was 42 mph (68 km/h), and that the distance is approximately 1900 miles (3060 km), we might crudely calculate the energy used in the Trans-Australian race as: 30 watt-hours/mile × 1900 miles = 57 kWh. At a typical U.S. electric energy cost of 10 cents per kilowatt-hour, that is approximately $5.70. For the GM *Impact*, which consumes 113 watt-hours per mile at 50 mph (80 km/h), the energy use would be approximately 214 kWh at a cost of $21.40.

Characteristics of *Sunraycer* are available from the Society of Automotive Engineers,[6] as well as other references to solar-powered vehicles.[2,8]

The next chapter discusses photovoltaic cells.

# References

1. *New York Times,* 25 January 1994. See also: Kaufman, Leslie, "Stempel's Revolution," *Newsweek*, 17 July 1995, p. 41.
2. Fujinaka, Masaharu, "Future Vehicles Will Run with Solar Energy," SAE Paper No. 891661, Conference on Future Transportation Technology, Vancouver, British Columbia, Canada, 7–10 August 1989, Society of Automotive Engineers, Warrendale, PA, 1989.
3. Wakefield, Ernest H., *History of the Electric Automobile: Battery-Only Powered Cars*, Society of Automotive Engineers, Warrendale, PA, 1994, p. 541.
4. Tuckey, Bill, *Sunraycer's Solar Saga,* BFT Publishing Group, Gordon, New South Wales, Australia, 1987.
5. Tholstrup, Hans, and Perkins, Larry, "Across Australia by Sun Power," *National Geographic,* **164**, 5, November 1983, pp. 600–607.

6.  MacCready, Paul B., *GM Sunraycer Case History,* Society of Automotive Engineers, Warrendale, PA, 1990, p. 12.
7.  *Michigan Today,* The University of Michigan, Ann Arbor, MI, Fall 1990.
8.  Kumagai, Naotake, and Tatemoto, Minoru, "Application of Solar Cells to the Automobile," SAE Paper No. 891696, Conference on Future Transportation Technology, Vancouver, British Columbia, Canada, 7–10 August 1989, Society of Automotive Engineers, Warrendale, PA, 1989.

# CHAPTER 10

# Photovoltaic Cells
# and Their Status

With the early efforts expended by searchers for an electric or multi-powered electric vehicle to supersede, even to a tiny degree, the truly remarkable gasoline-powered automobile, a possible partial solution is now emerging. This solution, while presently no larger than "a cloud the size of a man's hand rising from the sea," has possibilities.[1] This emerging icon is the solar-dominant and the solar-assist electric automobiles, the former sometimes called a solarmobile race car. In both vehicles, an onboard battery is constantly being electrically charged with power from the sun, a technology made cost efficient by increasing the efficiency while decreasing the cost of converting sunlight to electricity. Before reading the specifications of three well-designed (at the time) solar-dominant automobiles cited in Chapter 11, let us review the history of their onboard solar generators of electricity. This technology almost certainly will gain an ever-growing part of a nation's infrastructure. We should be aware of the history of this technology and know of its descent from pure science.

The chosen method for converting solar energy into electricity is by means of the photovoltaic or photoelectric effect. The photoelectric effect is defined as "the emission of an electron from a surface as the surface absorbs a photon of electromagnetic radiation (sun or daylight, in this book's interpretation). Electrons so emitted are termed photoelectrons."[2] Such a source of electricity is called a solar cell or a photovoltaic cell. When many of these cells are properly connected for increased power output, they are described as a solar module or a solar array. Although the propulsion forces for steam-, electric-, and gasoline-powered automobiles are probably better known to the reader, the development of photovoltaic cells is outlined here. This is possibly a more arcane subject but one essential to powering the solar-electric motorcar.

# History of the Photovoltaic Cell

The Frenchman Alexandre Edmund Becquerel, in noting what is now known as the photoelectric effect, observed that when one plate of a battery was illuminated, its voltage was altered.[3] His observation initiated a 65-year search to determine the mechanism of the effect. In 1873, Willoughby Smith noted that the electrical resistance of a selenium rod decreased when exposed to sunlight.[4] Later, this principal was used as a means of measuring light in quantitative terms. Heinrich Hertz, who early transmitted electric energy without wires, a concept later exploited as "wireless" (radio) by Guglielmo Marconi in 1887, discovered the photoemission properties of some metals when exposed to ultraviolet light but only from the negative terminal.[5] Three years later, in 1890, Johann Elster and Hans Geitel found that when alkaline metals, such as an amalgam of sodium and potassium, were enclosed in an evacuated glass envelope with suitable electrodes, they emitted negative particles when exposed to *visible light*. This tube was the first photocell.[6]

In 1899, the well-known physicist, J.J. Thompson, found that the emitted particles were identical in charge and mass, regardless of the type of metal cathodes used in a partially evacuated glass enclosure through which electricity was being passed. He called these particles *electrons*.[7] The next year (1900), Philipp Lenard used an evacuated glass envelope that was suitably equipped with electrodes and placed near a magnetic field, the intensity of which could be varied. Lenard unequivocally reported that the particles emitted by an illuminated surface were negative electrons.[8] The long search to explain the phenomenon of the photoelectric effect ended when Albert Einstein advanced the fundamental theory of photoelectric emission in 1905,[9] based on Max Planck's quantum theory of 1900. Planck proposed a value, now called Planck's constant. "The fundamental constant of quantum mechanics expressing the ratio of the energy of one quantum of radiation to the frequency of the radiation is approximately equal to $6.642 \times 10^{-27}$ erg-seconds"[10]:

$$\frac{\text{Energy}}{\text{Frequency}} = 6.642 \times 10^{-27} \text{erg} - \text{seconds}$$

Now that the phenomenon was explained, Elster and Geitel, in continuing their work in 1913, found that hydride crystals of sodium and potassium were approximately 100 times more sensitive to visible light than a pure metal. They also discovered the proportionality of light intensity and the emitted photocurrent from the surface. Their phototube, with hydrogenated alkali metal cathodes, was the beginning of modern phototube development.[11]

In all this experimentation, a consensus emerged. This consensus was that two general laws are an integral part of photoelectricity:

1. The number of electrons released per unit time at a photoelectric surface is directly proportional to the intensity of the incident light.

2. The maximum energy of the electrons released at a photoelectric surface is independent of the intensity of the incident light but increases linearly with the frequency of the light.[12]

Experiments continued in photoelectricity, particularly because the method was so fertile in explaining atomic structure, a subject studied intensely in the decades immediately before and after the turn of the century. With the development of solid-state electronics, Paul Rappaport in 1954 suggested using silicon-based semiconductors as sources of photoelectric power because they are more rugged than phototubes.[13] Later in the same year, D.M. Chapin, C.S. Fuller, and G.L. Pearson were the first to use a silicon cell in converting sunlight to electricity. Such cells are relatively inexpensive, have a reasonably high manufacturing yield, and possess an efficiency for converting sunlight to electricity of approximately 12 to 22%, roughly the efficiency of an automobile engine.[14] Figure 10.1 shows a cross section of a silicon solar cell.[15] Figure 10.2 is a schematic of the photoelectric effect.[16] Figure 10.3 demonstrates energy flow in a crystalline-silicon cell to a load. Finally, Table 10.1 lists trends and technical milestones of silicon photovoltaic cells from 1980 through 1994 in 1994 American dollars.[a] Another authority predicts solar-based electric energy of $0.07/kWh in the year 2000,[17] a cost said to be competitive with utility power.

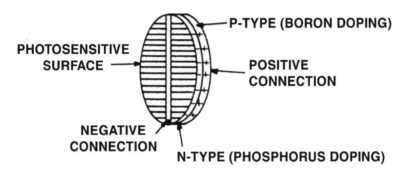

*Figure 10.1. A drawing of a photovoltaic cell. The base material at the right of the figure is made of P-type silicon. After the P-type material is formed by adding boron doping, a thin layer of N-type material is formed by changing the doping to phosphorus. The N-type layer is very thin. Although it appears opaque, sunlight will penetrate it rather deeply. In fact, the light penetrates the junction between the N-type and P-type materials. This place is where the hole-electron pairs are formed. The electric field that exists at the junction will prevent the holes and electrons from recombining, with the result that we can now use the device as a source of energy. The N-type material will be the negative pole, and the P-type material will be the positive pole of what amounts to a little electric generator that obtains its energy from sunlight. (Ref. 15)*

---

[a] Production and cost of silicon solar cells have moved in opposite directions in the past decade, and the trend is projected to continue. A module is a sealed panel containing a number of solar cells interconnected at the factory; several modules constitute an array. (Strategies Unlimited, 201 San Antonio Circle, Suite 205, Mountain View, CA 94040)

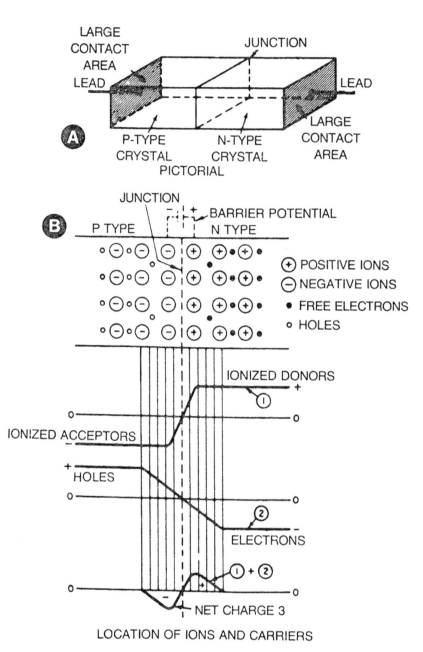

*Figure 10.2. When a photon penetrates a semiconductor material such as silicon, an electron will be forced out of its place in the crystal structure. This displacement forms a hole-electron pair. The electron has a negative charge, and the hole that it leaves will have a positive charge. This phenomenon is the photoconductive effect. The electrons freed by photons of appropriate energy become more energetic and move from the valence, or bound state, to the conduction band. If the photoconductive effect occurs near the electric field of a cell, the photogenerated hole-and-electron pair are separated and move to opposite ends of the cell, becoming a part of the electric current. (Ref. 16)*

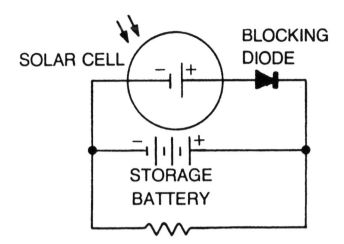

*Figure 10.3. Energy flow in a silicon solar cell to a load. The diode prevents battery current from flowing through the solar cells when its output is zero. (Solarex Corporation)*

**TABLE 10.1**
**WORLDWIDE SHIPMENTS BY REGION, 1980–1994[1]**

| Year | United States kWp | %T[2] | Europe kWp | %T[2] | Japan kWp | %T[2] | ROW kWp | %T[2] | Total $M | kWp | ASP[3] $/Wp |
|---|---|---|---|---|---|---|---|---|---|---|---|
| 1980 | 2,876 | 86 | 374 | 11 | 68 | 2 | 18 | 1 | 34.3 | 3,330 | 10.30 |
| 1981 | 3,960 | 75 | 850 | 16 | 435 | 8 | 50 | 1 | 50.3 | 5,295 | 9.50 |
| 1982 | 4,630 | 60 | 1,540 | 20 | 1,400 | 18 | 130 | 2 | 73.8 | 7,700 | 9.59 |
| 1983 | 9,320 | 64 | 2,150 | 15 | 2,760 | 19 | 300 | 2 | 114.1 | 14,530 | 7.85 |
| 1984 | 8,400 | 48 | 2,900 | 17 | 5,300 | 30 | 800 | 5 | 138.2 | 17,400 | 8.04 |
| 1985 | 7,100 | 38 | 3,400 | 18 | 6,400 | 34 | 1,800 | 10 | 130.3 | 18,700 | 7.03 |
| 1986 | 6,400 | 30 | 3,500 | 17 | 8,200 | 40 | 2,900 | 14 | 135.3 | 21,000 | 6.44 |
| 1987 | 7,400 | 30 | 4,450 | 18 | 8,800 | 35 | 4,200 | 17 | 148.9 | 24,900 | 5.99 |
| 1988 | 10,700 | 34 | 6,400 | 20 | 9,100 | 29 | 5,300 | 17 | 185.8 | 31,500 | 5.90 |
| 1989 | 13,650 | 36 | 6,800 | 18 | 10,800 | 29 | 6,600 | 17 | 231.4 | 37,850 | 6.10 |
| 1990 | 15,200 | 36 | 9,050 | 21 | 13,000 | 30 | 5,400 | 13 | 271.6 | 42,650 | 6.37 |
| 1991 | 17,000 | 35 | 10,500 | 22 | 14,900 | 31 | 5,800 | 12 | 296.7 | 48,200 | 6.16 |
| 1992 | 18,600 | 40 | 13,950 | 26 | 13,750 | 25 | 7,800 | 14 | 319.7 | 54,100 | 5.91 |
| 1993 | 22,300 | 40 | 13,700 | 25 | 12,500 | 22 | 7,200 | 13 | 304.8 | 55,700 | 5.47 |
| 1994 | 24,550 | 40 | 14,700 | 24 | 12,850 | 21 | 8,900 | 15 | 328.7 | 61,000 | 5.39 |

(1) Module and cell revenue only; current dollars; market share based on kWp shipments.

(2) Percentage of total world shipments in kWp.

(3) Module average selling price, current dollars, FOB factory.

Source: Strategies Unlimited.

In referring to energy losses in a photovoltaic cell, Yashihiro Hamakawa[18] writes:

> Energy flow in a crystalline-silicon solar cell utilizes in principle about 44% of the solar energy to which the cell is sensitive. About 16% of the energy is lost by various processes within the cell, leaving a theoretical limit of efficiency of 28%. Further losses reduce the actual efficiency of the cell in converting sunlight into electricity to between 14 and 22%.

Even at present costs, these daylight-to-electricity converters were soon being placed in service as a source of electrical energy in areas where electric utility lines were absent: isolated ship-channel navigation lights, railroad crossing warnings, and later, as satellite power sources. Reflecting on the popularity of silicon-based solar cells, almost all solar-electric cars shown in the following illustrations bear silicon-based solar cells in their solar arrays largely because of their moderate cost-per-watt (CPW) factor. *The Solarex Guide to Solar Electricity,*[15,16] provides practical information concerning silicon solar cells.

# Gallium-Arsenide Solar Cells

With the desire to increase the conversion efficiency of sunlight, other semiconductors are being evaluated. Now entering service are gallium-arsenide (GaAs) solar cells. How recently this has occurred is reflected in A. Shibatomi's 1987 paper on their improvement.[19] Although considerably more expensive than silicon-based cells, the former have conversion efficiencies of 18 to 22%.[14] The *Sunraycer* entered by General Motors in the Pentax World Solar Challenge Race is described in Chapter 13. The race vehicle, the second of the two vehicles built, bears both silicon-based and GaAs cells.[20] Inasmuch as the rules of the race limited the solar active area to eight square meters, any solarmobile with an array bearing a sizable portion of GaAs cells as compared to silicon-based cells would naturally bear a built-in advantage over a contestant that used only silicon-based solar cells. The hazard of arsenic handling has been emphasized to me by Norman N. Lichtin, Professor of Chemistry at Boston University. Therefore, safety rules should be followed carefully.[21]

Following this brief history of photovoltaic cells, the next chapter presents maps of sun areas of the world.

# Note 1—More on Gallium-Arsenide Solar Cells

In reviewing the sections of this book that deal with solar power, William B. Rever III, Product Manager in the Crystalline Division of the Solarex Corporation, writes[22]:

> Since GaAs (gallium-arsenide) solar cells cost $10^2$ to $10^3$ as much as terrestrial Si (silicon) solar cells, the only possible applications are in very high value situations such as military satellites. Although some of these costs could be reduced with true mass production,

there are many problems inherent in large scale manufacturing of GaAs. First, it is very toxic and cannot be handled like Si. All of the wafer manufacturing operations (coating, sawing, etc.) must be highly protected to avoid poisoning workers. Each of these operations also produces large quantities of very toxic waste. If the use of these materials ever becomes as widespread as we hope PV (photovoltaics) will be, there will probably be supply problems with such relatively exotic materials. Finally, the widespread deployment of such toxic substances could create numerous public safety hazards: vehicles crashing and burning, disposal of vehicles, maintenance, etc. For these reasons, I think it will be unlikely (that GaAs photovoltaics will be widely used in transportation).

# Note 2—Economic Considerations

When I was studying electrical engineering at the University of Michigan in the 1930s, professors of power engineering indicated central station power output capital cost was approximately $100 to $150 per kilowatt from coal-fired central stations. A recent efficiently built nuclear station had an equivalent cost of $2600 per kilowatt, allowing for inflation change in the value of the dollar. Hamakawa's figure of $1000 per kilowatt from silicon cells in the year 2000[17] does have economic and environmental interest because its power generation is clean; furthermore, solar power is daytime power, which is the period of high usage.

# Note 3—Solar Cells in Boats

I have been astonished by the application of solar cells to powering electric boats where, similar to solar-battery powered cars, solar-battery powered boats are surprisingly popular. Even international regattas exist for this type of personal travel, although battery-only powered boats have reached speeds greater than 56 mph (93 km/h).

# Note 4—The Wave Theory of Light

The conflict over the nature of light—whether it is propagated by means of corpuscles or in wave form—led strangely to the deciphering of the ancient language of the Pharaohs. The tale, taken in part from *Electric and Magnetic Fields* by Stephen S. Attwood,[23] is related in the Introduction of my book, *The Lighthouse That Wanted To Stay Lit*.[24]

The person who most influenced lighthouse design worldwide was almost surely Augustin Jean Fresnel (1788–1827). Fresnel was born at Broglie, near Caen in French Normandy, to middle-class parents. After attending the well-regulated Caen school, his parents sent him to the *École Polytechnique* and then to the *École des Ponts et Chaussées*. Thereafter, he was employed by the royal state in designing and building bridges and roads. Although lenses for light had been known since the year 1039, Galileo Galilei (1564–1642) had used a lens-equipped telescope in discovering several of the moons of Jupiter. However, little was known about the nature of light. In his free time, Fresnel pursued optics, a science dominated by Sir Isaac Newton (1642–1727) and his corpuscular theory of light. Newton was one of the greatest scientists who ever lived and had such prestige that his theory of light was believed everywhere for nearly 150 years.

With Napoleon's triumphant return from the island of Elba after his first banishment, Fresnel, a supporter of Louis XVIII, opposed him and later lost his state position under the Empire. In this enforced freedom, Fresnel conducted experiments in optics, particularly on the diffraction of light. His intense work in optics began in 1815, after the Battle of Waterloo. With the resulting restoration of royalty under Louis XVIII, Fresnel again was given his state position. Independently he concluded that light was propagated as a wave form. His friend, Dominique François Arago, called his attention to the papers of Dr. Thomas Young, an English physicist and physician, with whom Fresnel was unacquainted because of the blockade imposed by the Napoleonic wars. Young, with his experiments, also was persuaded to the wave theory. Indeed, Young had been so severely attacked by fellow scientists for his beliefs that he abandoned the study of optics and devoted himself to hieroglyphic research. In this task, he successfully deciphered the Rosetta Stone, which had been found in Egypt by an officer from Napoleon's forces near the original site of Pharos (lighthouse). Thus, the language used by the Pharaohs became known.

Fresnel brought to his research "an ingenious mind, deft hands, and the discipline of an excellent scientific education" and succeeded in winning an ever increasing number of scientists to the wave theory of light.[24] Toward the end of his life, Fresnel devoted his considerable knowledge in optics to the French Bureau of Lighthouses. His optic designs for lighthouses affected the entire world. He died of tuberculosis in his home in Broglie on 14 July 1827. His extraordinary creative life was nine years.

# References

1. King James Bible, 1. Kings 18:44.
2. Weast, Robert C., ed., *CRC Handbook of Chemistry & Physics*, CRC Press, Boca Raton, FL, 1987, p. F-97.
3. Becquerel, E., *Studies of the Effect of Actinic Radiation of Sunlight by Means of Electric Currents, Compt. rend.,* Vol. 9, 1839, pp. 145–149.
4. Smith, W., "Effect of Light on Selenium During the Passage of an Electric Current," *Am. J. Sci.,* Vol. 9, 1873, p. 301.
5. Hertz, H., "Ultra-Violet Light and Electric Discharges," *Ann. Physik*, Vol. 31, 1887, pp. 983–1000.
6. Elster, J., and Geitel, H., "The Use of Sodium Amalgam in Photoelectric Experiments," *Ann. Physik*, Vol. 41, 1890, pp. 161–165.
7. Thompson, J.J., "On the Masses of Ions in Gases at Low Pressure," *Phil. Mag.*, Vol. 48, 1899, pp. 547–567.

8.  Lenard, P., "Production of Cathode Rays by Ultra-Violet Light," *Ann. Physik*, Vol. 2, 1900, pp. 359–75.
9.  Einstein, A., *Ann. d. Phys.*, Vol. 17, 1905, p. 132.
10. Planck, M., "On the Theory of the Law of Energy Distribution in the Normal Spectrum," *Verhandl. deut. Physik, Ges.*, Vol. 2, 1900, pp. 237–245.
11. Elster, J., and Geitel, H., "Proportionality of Light Intensity and Photocurrent in Alkali Metal Cells," *Physik. Z.*, Vol. 14, 1913, pp. 741-752.
12. Zworykin, V.K., and Ramberg, E.G., *Photoelectricity and Its Application*, John Wiley, New York, 1949.
13. Moss, T.S., Burrell, G.S., and Ellis, B., *Semiconductor Opto-Electronics*, Wiley, New York, 1973, pp. 313–314.
14. Glasstone, Samuel, *Energy Deskbook*, Van Nostrand Reinhold, New York, 1983, pp. 348–356.
15. Robertson, Edward, *et al., The Solarex Guide to Solar Electricity*, Tab Books, Blue Ridge Summit, PA, 1983, p. 37.
16. *Ibid.*, 21–24.
17. Hamakawa, Yashihiro, "Recent Progress in Solar Photovoltaic and Its Contribution to Environmental Issues," *Business Japan*, September 1990, pp. 26–30.
18. Hamakawa, Yashihiro, "Photovoltaic Power," *Scientific American*, April 1987, p. 90.
19. Shibatomi, A., "Surface Passivation for GaAs," Moore, D.F., ed., *Solid State Devices 1986*, IOP Publishing Ltd., Bristol, England, 1987, pp. 185–201.
20. Brooks, Alec N., Car Development Manager of *Sunraycer*, AeroVironment Inc., personal communication, 25 March 1988.
21. Lichtin, Norman N., Professor of Chemistry, Boston University, personal communication, August 1993.
22. Rever, William B. III, Solarex Corporation, personal communication, 7 August 1990.
23. Attwood, Stephen S., *Electric and Magnetic Fields*, Wiley, New York, 1949.
24. Wakefield, Ernest H., *The Lighthouse That Wanted To Stay Lit*, Honors Press, Evanston, IL, 1992, pp. xxx-xxxiii.

# CHAPTER 11

# Sun Areas of the World and the Atmosphere

## Geographic Considerations in Solar-Electric Vehicle Design

In designing a solar-electric powered vehicle, we begin by knowing the value of the so-called *solar constant*. This quantity is the amount of radiation from the sun received by an imaginary surface normal to its rays located at the top of the atmosphere. Its value is approximately 1.35 kW/m$^2$ (1.2 yd$^2$).[1] At ground level, this irradiation is reduced to approximately 1.0 kW/m$^2$, the result of factors such as air scattering, water, and dust absorption.[2]

Cities in the high sun regions of the world particularly offer an opportunity for the solar-electric cars described here. Figure 11.1 demonstrates worldwide distribution of the sun in watt-hours/week (yearly average) for a particular Solarex module.[3] The important concept to grasp is a solar-electric vehicle would have the greatest range between battery electrical recharges in regions where the iso-energy lines seen in the figure have the highest number. Note the higher numbers in the southwestern part of the United States. Alphabetized by states, Table 11.1 lists some of America's principal cities, showing average percent of possible sunshine received by them per year.[a] The table simply confirms Figure 11.1 in greater detail for the United States.

---

[a]For a complete list, see the World Wide Web site at http://info.abrfc.noaa.gov/wfodocs/sunshine.html/.

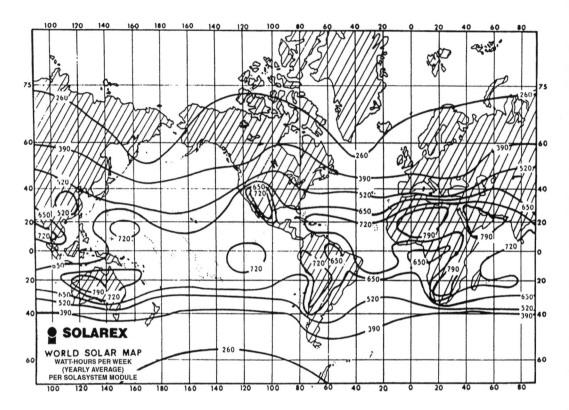

*Figure 11.1. Worldwide distribution, average watt-hours/week (yearly average) per system module. (Solarex Corporation. Ref. 3)*

For example, Los Angeles is seen not only to be in a region of high sunlight-hours, but the city's atmosphere is one that requires close attention. Hence, this region might be a receptive venue for offering solar-electric vehicles. However, the average annual peak-sun hours in the balance of the United States (Figure 11.2) are not greatly lower than for the Los Angeles region.[4] Finally, Figure 11.3 provides data for peak sun-hours per day for a four-week period of mid-winter in America.[5]

This power is the maximum available for a ground-based solar array, and Figure 11.4 traces this flow of power in the drive system of a solarmobile. In the figure, the solar cell efficiency is given as 12%, indicating that Professor Harry West of Massachusetts Institute of Technology, who made the drawing, was thinking in terms of silicon-based solar cells. Figure 11.5 shows a qualitative energy loss distribution in percent versus speed for any electric vehicle and specifically for a van.

## TABLE 11.1
## AVERAGE PERCENT POSSIBLE SUNSHINE RECEIVED IN SELECTED CITIES

## Average Percent Possible Sunshine

| DATA THROUGH 1986 | YRS | J | F | M | A | M | J | J | A | S | O | N | D | A |
|---|---|---|---|---|---|---|---|---|---|---|---|---|---|---|
| BIRMINGHM C.O.,AL | 7 | 48 | 50 | 63 | 64 | 64 | 66 | 62 | 64 | 60 | 58 | 46 | 53 | 58 |
| BIRMINGHM AP, AL | 34 | 42 | 50 | 55 | 63 | 66 | 65 | 59 | 63 | 61 | 66 | 55 | 46 | 58 |
| MONTGOMERY, AL | 36 | 48 | 54 | 59 | 65 | 65 | 64 | 62 | 64 | 62 | 65 | 55 | 50 | 59 |
| ANCHORAGE, AK | 33 | 39 | 45 | 54 | 53 | 52 | 48 | 44 | 42 | 42 | 39 | 36 | 32 | 44 |
| JUNEAU, AK | 33 | 32 | 32 | 37 | 39 | 39 | 34 | 31 | 32 | 26 | 19 | 23 | 20 | 30 |
| NOME, AK | 38 | 40 | 54 | 54 | 53 | 50 | 41 | 37 | 30 | 35 | 35 | 31 | 35 | 41 |
| FLAGSTAFF, AZ | 8 | 77 | 72 | 72 | 81 | 88 | 84 | 70 | 73 | 78 | 76 | 73 | 71 | 76 |
| PHOENIX, AZ | 91 | 78 | 80 | 83 | 88 | 93 | 94 | 85 | 85 | 89 | 88 | 83 | 77 | 85 |
| TUCSON, AZ | 39 | 81 | 84 | 86 | 91 | 94 | 93 | 78 | 81 | 87 | 88 | 85 | 80 | 86 |
| YUMA, AZ | 36 | 84 | 87 | 90 | 94 | 96 | 97 | 90 | 91 | 93 | 92 | 87 | 82 | 90 |
| FORT SMITH, AR | 41 | 51 | 55 | 57 | 59 | 62 | 69 | 73 | 72 | 66 | 64 | 55 | 51 | 61 |
| LITTLE ROCK, AR | 32 | 46 | 54 | 57 | 62 | 68 | 73 | 71 | 73 | 68 | 69 | 56 | 48 | 62 |
| N LITTLE ROCK, AR | 9 | 67 | 66 | 72 | 75 | 76 | 82 | 85 | 80 | 79 | 70 | 57 | 58 | 72 |
| EUREKA, CA | 76 | 42 | 45 | 52 | 57 | 57 | 58 | 54 | 49 | 54 | 49 | 43 | 40 | 50 |
| FRESNO, CA | 37 | 47 | 65 | 77 | 85 | 90 | 94 | 96 | 96 | 93 | 88 | 65 | 45 | 78 |
| L ANGELES C.O.,CA | 32 | 69 | 72 | 73 | 70 | 66 | 65 | 82 | 83 | 79 | 73 | 74 | 71 | 73 |
| RED BLUFF, CA | 42 | 54 | 64 | 70 | 81 | 86 | 89 | 96 | 94 | 92 | 81 | 61 | 53 | 77 |
| SACRAMENTO, CA | 38 | 44 | 62 | 72 | 81 | 89 | 93 | 97 | 96 | 93 | 85 | 63 | 46 | 77 |
| SAN DIEGO, CA | 46 | 72 | 72 | 70 | 67 | 58 | 57 | 69 | 70 | 68 | 68 | 74 | 72 | 68 |
| SAN FRAN. C.O.,CA | 38 | 56 | 62 | 69 | 73 | 72 | 73 | 66 | 65 | 72 | 70 | 62 | 53 | 66 |
| DENVER, CO | 37 | 71 | 71 | 70 | 68 | 65 | 71 | 71 | 72 | 74 | 72 | 64 | 67 | 70 |
| GRAND JUNCTION,CO | 40 | 60 | 64 | 64 | 69 | 72 | 80 | 78 | 76 | 78 | 73 | 63 | 60 | 70 |
| PUEBLO, CO | 45 | 75 | 74 | 74 | 75 | 74 | 79 | 78 | 78 | 80 | 78 | 73 | 72 | 76 |
| HARTFORD, CT | 32 | 57 | 57 | 56 | 56 | 58 | 60 | 64 | 62 | 59 | 58 | 47 | 49 | 57 |
| WASH NAT'L AP, DC | 38 | 48 | 51 | 55 | 57 | 59 | 64 | 63 | 63 | 62 | 58 | 51 | 47 | 57 |
| APALACHICOLA, FL | 51 | 59 | 62 | 65 | 74 | 78 | 72 | 64 | 64 | 66 | 74 | 67 | 57 | 67 |
| JACKSONVILLE, FL | 35 | 58 | 62 | 68 | 72 | 70 | 63 | 62 | 60 | 56 | 58 | 60 | 56 | 62 |
| KEY WEST, FL | 27 | 73 | 76 | 83 | 84 | 81 | 74 | 76 | 75 | 71 | 70 | 70 | 70 | 75 |
| MIAMI, FL | 10 | 69 | 68 | 77 | 78 | 71 | 74 | 76 | 75 | 72 | 72 | 67 | 65 | 72 |
| PENSACOLA, FL | 5 | 48 | 53 | 61 | 63 | 67 | 67 | 57 | 58 | 60 | 71 | 64 | 49 | 60 |
| TAMPA, FL | 39 | 64 | 66 | 72 | 75 | 75 | 67 | 61 | 60 | 61 | 64 | 65 | 61 | 66 |
| ATLANTA, GA | 51 | 49 | 54 | 58 | 66 | 68 | 67 | 63 | 64 | 64 | 67 | 59 | 51 | 61 |
| MACON, GA | 38 | 55 | 60 | 64 | 70 | 71 | 70 | 65 | 70 | 65 | 69 | 63 | 57 | 65 |
| SAVANNAH, GA | 36 | 55 | 58 | 62 | 70 | 67 | 65 | 62 | 62 | 58 | 63 | 61 | 55 | 62 |
| HILO, HI | 36 | 47 | 45 | 40 | 35 | 37 | 43 | 42 | 42 | 44 | 39 | 35 | 38 | 41 |
| HONOLULU, HI | 34 | 62 | 64 | 68 | 67 | 69 | 70 | 73 | 75 | 75 | 68 | 61 | 59 | 68 |
| KAHULUI, HI | 24 | 65 | 65 | 64 | 63 | 69 | 73 | 72 | 72 | 75 | 70 | 65 | 64 | 68 |
| LIHUE, HI | 36 | 52 | 55 | 53 | 52 | 57 | 61 | 62 | 64 | 66 | 58 | 49 | 48 | 56 |
| BOISE, ID | 44 | 38 | 49 | 62 | 68 | 71 | 75 | 87 | 85 | 80 | 68 | 44 | 38 | 64 |
| POCATELLO, ID | 37 | 39 | 52 | 61 | 65 | 67 | 74 | 82 | 80 | 78 | 70 | 47 | 39 | 63 |
| CAIRO, IL | 44 | 45 | 51 | 55 | 62 | 65 | 72 | 74 | 75 | 69 | 67 | 51 | 44 | 61 |
| CHICAGO, IL | 6 | 46 | 42 | 51 | 50 | 58 | 68 | 69 | 66 | 56 | 48 | 37 | 42 | 53 |
| MOLINE, IL | 43 | 48 | 49 | 50 | 53 | 56 | 62 | 68 | 66 | 63 | 58 | 42 | 40 | 55 |
| PEORIA, IL | 43 | 47 | 50 | 51 | 55 | 59 | 66 | 68 | 67 | 64 | 61 | 44 | 41 | 56 |
| SPRINGFIELD, IL | 38 | 48 | 51 | 51 | 56 | 64 | 68 | 72 | 71 | 68 | 62 | 48 | 42 | 58 |
| EVANSVILLE, IN | 46 | 43 | 48 | 55 | 59 | 65 | 72 | 74 | 75 | 70 | 65 | 48 | 41 | 60 |
| FORT WAYNE, IN | 40 | 45 | 50 | 53 | 59 | 66 | 72 | 74 | 73 | 67 | 61 | 41 | 38 | 58 |
| INDIANAPOLIS, IN | 42 | 41 | 49 | 50 | 54 | 60 | 66 | 67 | 69 | 66 | 61 | 42 | 38 | 55 |
| DES MOINES, IA | 36 | 52 | 54 | 54 | 55 | 61 | 68 | 73 | 70 | 65 | 61 | 49 | 49 | 59 |
| SIOUX CITY, IA | 46 | 58 | 57 | 58 | 60 | 63 | 68 | 74 | 71 | 66 | 63 | 52 | 51 | 62 |
| CONCORDIA, KS | 24 | 63 | 63 | 63 | 65 | 68 | 77 | 79 | 76 | 69 | 69 | 60 | 58 | 68 |
| DODGE CITY, KS | 44 | 67 | 64 | 64 | 67 | 68 | 75 | 79 | 77 | 74 | 72 | 65 | 64 | 70 |
| TOPEKA, KS | 37 | 56 | 54 | 55 | 56 | 59 | 64 | 70 | 69 | 64 | 62 | 53 | 51 | 59 |
| WICHITA, KS | 33 | 60 | 60 | 61 | 64 | 65 | 69 | 75 | 74 | 67 | 65 | 58 | 58 | 65 |
| GTR CINCINNATI AP | 3 | 40 | 38 | 45 | 57 | 55 | 65 | 68 | 65 | 68 | 47 | 32 | 34 | 51 |
| LOUISVILLE, KY | 39 | 42 | 47 | 50 | 55 | 61 | 66 | 66 | 67 | 65 | 61 | 46 | 40 | 56 |
| LAKE CHARLES, LA | 5 | 55 | 56 | 70 | 74 | 81 | 83 | 85 | 84 | 81 | 68 | 54 | 53 | 70 |
| NEW ORLEANS, LA | 13 | 48 | 52 | 59 | 64 | 62 | 67 | 62 | 61 | 64 | 66 | 53 | 52 | 59 |
| SHREVEPORT, LA | 34 | 49 | 55 | 57 | 57 | 63 | 70 | 73 | 72 | 68 | 67 | 58 | 53 | 62 |
| PORTLAND, ME | 46 | 56 | 59 | 56 | 55 | 55 | 59 | 64 | 63 | 62 | 58 | 48 | 52 | 57 |
| BALTIMORE, MD | 36 | 51 | 55 | 56 | 56 | 57 | 62 | 65 | 62 | 60 | 58 | 50 | 48 | 57 |
| BLUE HILL, MA | 100 | 46 | 50 | 48 | 50 | 52 | 55 | 57 | 58 | 56 | 55 | 47 | 46 | 52 |
| BOSTON, MA | 51 | 53 | 56 | 57 | 56 | 58 | 63 | 66 | 65 | 64 | 60 | 50 | 52 | 58 |
| ALPENA, MI | 27 | 38 | 44 | 52 | 54 | 59 | 63 | 67 | 60 | 51 | 44 | 30 | 27 | 49 |
| DETROIT, MI | 21 | 40 | 45 | 51 | 55 | 61 | 66 | 70 | 69 | 61 | 50 | 35 | 29 | 53 |

197

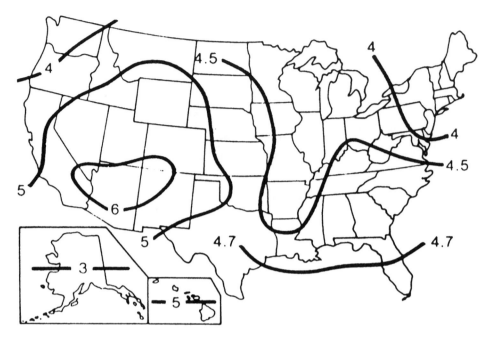

*Figure 11.2. Yearly peak sun-hours per day in America. (Solarex Corporation. Ref. 4)*

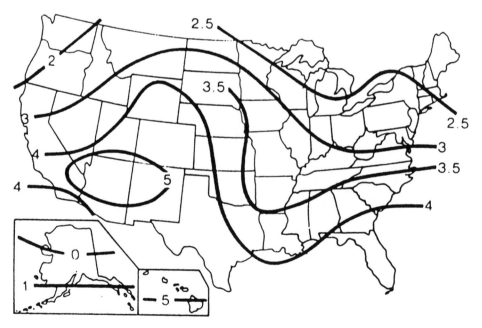

*Figure 11.3. Peak sun-hours per day for a four-week period from 7 December to 4 January. (Solarex Corporation. Ref. 5)*

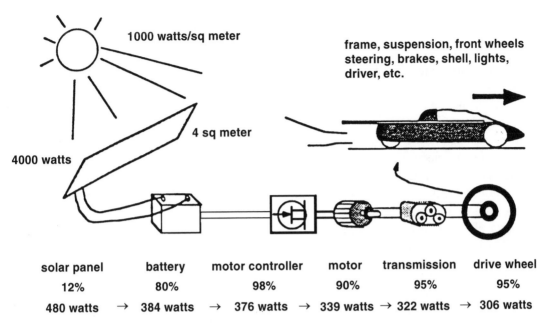

*Figure 11.4. Demonstrating a flow of power in a solar-electric vehicle.*
*(Professor Harry West, Massachusetts Institute of Technology)*

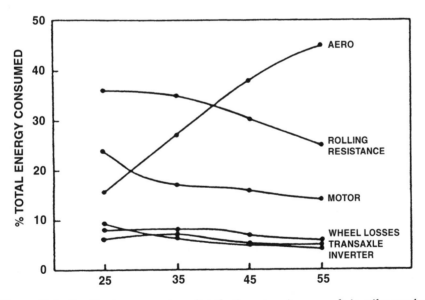

*Figure 11.5. Qualitative energy loss distribution at various speeds in miles per hour*
*(for an electric van ETX-1). (Ref. 6)*

199

# Solar-Assist and Solar-Dominant Motorcars

At this genesis stage of solar-electric vehicle development, the designer has a constriction of far greater importance than do his peers who are responsible for internal combustion vehicles. Whereas the latter can exercise wide latitude in the power of the vehicle engine, the maximum power from the sun is absolutely limited to the value previously cited, a relatively small number when speaking of powering conventional motorcars. Two types of sun-using vehicles are being developed now : 1) the *solar-assist* electric car, as shown in Figure 14.5, and 2) the *solar-dominant* electric car, as sketched in Figure 11.4. With a severe power restriction, particularly in the latter, the solar-electric designer is encouraged to incorporate the following in both classes of cars: 1) a lightweight vehicle with a modest electric battery complement, 2) a streamlined vehicle, 3) a vehicle with minimum rolling resistance, 4) a vehicle with a highly efficient solar array, with realization the sun is an angle-changing body, and 5) a vehicle with a highly efficient motor-drive system.

Figure 14.5 (courtesy of Douglas Cobb of Solar Car Corporation) assures us that a substantial stand of batteries is aboard the pick-up, with solar-assist cells fastened above the batteries, and on the forward hood. In contrast, the *Sunraycer,* is a *solar-dominant* car (described in Chapter 12) capable of achieving high speeds on solar power alone. On it, 360 solar cells in a string yield approximately 150 volts, and 20 strings in parallel provide an array capable of yielding 1500 watts from 8 m$^2$ of silicon and gallium-arsenide solar cells. As the sun wanes from its zenith, the voltage remains the same but the current decreases. This lessens the available power, the product of voltage and current, for the electric drive motor, if of the d-c type. An onboard chemical battery is seen as essential for solar-cell balancing and is especially invaluable for a solar-dominant car in providing additional electric power to the motor for short bursts of increased speed while passing another car, in providing power for accessories, or in providing emergency power when required (i.e., in hill climbing).

In any city, if an average driver's commute distance is 20 to 30 miles (32 to 48 km), that driver will want some kind of electric outlet at his parking lot, as shown in Appendix A. Using the vehicle characteristics of the General Motors electric *Impact I,* a putative *Impact II* (bearing 4 m$^2$ of solar cells), equipped with a 220-volt, 20-ampere onboard charger, the vehicle batteries would absorb from the outlet approximately 4.4 kWh in one hour, an amount of energy substantially greater than the energy consumed on many home-to-office commutes. In less than one hour on electrical charge, the batteries could be brought to full charge. Figure 11.6 demonstrates the calculated extra mileage that might be expected with an *Impact* equipped with, respectively, 4 m$^2$ of silicon-based and gallium-arsenide-based solar-assist cells assuming an energy consumption of 113 watt-hours/mile (continuous driving at 55 mph [89 km/h]). If half the area were equipped with cells, divide the extra mileage by two.

In Chicago, I drove electric cars in snow and in low-temperature conditions. Because of their battery load, the relatively heavy electric cars handled better in snow than lightweight vehicles. What about the effect of temperature on the batteries? If the batteries have been "soaking" in an ambient low temperature, the range of the vehicle may be appreciably reduced. I have

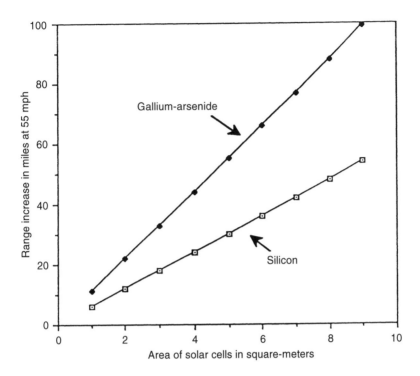

*Figure 11.6. Extra miles of range from different types of solar-assist on a General Motors EV1 type vehicle, assuming average peak-sun hours in Los Angeles and energy consumption of 113 watt-hours/mile.*

driven with batteries that had a thin layer of ice on the electrolyte. The best advice under such cold conditions is either to garage the electric car or, if the vehicle is not housed, to allow a trickle charge to flow into the battery, a natural event in many charging circuits when the batteries are fully charged. Suitably placed photovoltaic cells also can provide a trickle battery charge.

In vehicle operation, batteries heat up due to motor-current flowing through them. This phenomenon is known as $I^2R$ (pronounced "I squared R") heat loss. Because of the relative high voltage (320 volts) of operation of the *Impact* (EV1) drive system, this $I^2R$ loss is more moderate than in electric cars operating at a lower voltage. On the other hand, the battery serviceperson for the *Impact* must be more safety conscious because of the high voltage of the battery. If shorted by wrench or human hand, this battery (or any battery) could deliver instantly a powerful amount of energy! The importance of this battery heating on vehicle operation was discovered by the two college participants, Massachusetts Institute of Technology and California Institute of Technology, in their race across America in 1968 in electric cars. According to Dr. Victor Wouk[7]:

> MIT had to put ice on the bus-bars that connected the Ni/Cd cells during charging. The cubes melted fast during running. Cal Tech put ice water into the case that held the batteries, during charging. The water was siphoned off during running to reduce the

201

weight. If not for a suggestion that I had made to Wally Rippel a few days before the race, and which he agreed made sense, he would never have finished. He had the batteries in a 'solid block.' I told him to separate each module an inch or so, for cooling. It worked.

I have found that a slight mental adjustment must be made when driving an electric vehicle. I reached this conclusion after having introduced possibly as many as 100 persons to driving electric cars. Most of these novices were mature electric utility engineers. As stated by Cholly Knickerbocker in Chapter 4 of *History of the Electric Automobile: Battery-Only Powered Cars*, when he first rode in an electric taxi in New York in 1895, having always previously used a hansom cab drawn by a horse: "I don't regret the experiment."[8] Chances are, you won't either!

The next chapter discusses the history of solar-powered vehicles.

# Notes

For the scholar who has access to a large library and is interested in climatology and sunshine, two excellent books are as follows:

- Brysen, Reid A., and Hare, F. Kenneth, eds., *World Survey of Climatology, V. II, Climate of North America*, Elsevier Publishing Company, Amsterdam, 1974, pp. 237 and 238.

- Landsberg, H.E., Lippman, H., and Troll, C., *World Map of Climatology*, 2nd edition, Springer-Verlag, Berlin, 1965, pp. 3 and 4 and Index maps.

# References

1.  Weast, Robert C., ed., *CRC Handbook of Chemistry and Physics*, CRC Press, Boca Raton, FL, 1987, p. F-97.
2.  Omann, Henry, and Gelzer, Joseph W., "Solar Cells and Arrays," Considine, Douglas M., ed., *Energy Technology Handbook*, McGraw-Hill, New York, 1977, p. 6-56.
3.  Robertson, Ed, ed., *The Solarex Guide to Solar Electricity*, Tab Books, Blue Ridge Summit, PA, 1983, p. 64.
4.  *Ibid.*, p. 61.
5.  *Ibid.*, p. 63.
6.  Fenton, J., and Patil, P.B., "Advanced Electric Vehicle Powertrain (ETX-I) Performance: Vehicle Testing," Eighth International Electric Vehicle Symposium, Washington, DC, 1986, p. 500.
7.  Wouk, Victor, personal communication, 2 April 1994.
8.  Wakefield, Ernest H., *History of the Electric Automobile: Battery-Only Powered Cars*, Society of Automotive Engineers, Warrendale, PA, 1994, p. 47.

# CHAPTER 12

# History of Solar-Electric Vehicles

## Background

Vehicles of today follow several earlier modes: electric, steam, and the spark-ignition engine. The field was finally dominated by the last mode; thus, we might expect future diversification in solar-electric cars. However, this difference exists for solar power: the car designer initially pre-selected the final limited power sources, which were solar cells of some type, and an electric battery of a finite size. Of the battery, similar to the solar cell, several choices are possible, a subject discussed in *History of the Electric Automobile: Battery-Only Powered Cars.*[1]

This chapter includes examples in the gradual placing of solar cells on electric vehicles: first as a supplier for auxiliary needs of electricity; then as an important source of supplemental motive power[2]; then as the main source of drive-power at speeds of 16 mph (26 km/h) in 1982; the crossing of Australia by the *Quiet Achiever*; next, a 1982 cost study for making a trans-American solar-electric vehicle trip; the 1983 design plans for a solar-electric vehicle for the proposed 1992 Chicago's World Fair; then three illustrations from 1987; some world-class solar-electric vehicles capable of speeds of approximately 50 mph (80 km/h) while using only solar power; and, finally, several solar-assist automobiles.

Figure 12.1 illustrates a 1974 electric vehicle where the modest roof array of solar cells is limited to powering the accessories of the vehicle. The five cells reportedly yielded 600 milliamperes at 2.2 volts, a total of 1.3 watts.[3] In Ref. 2, written in 1977, solar cells are represented as the source of solar-assist power for an electric vehicle, as shown in Figure 12.2.[4] I wrote of the viability of energy from the sun, based on solar maps in Chapter 11, charging an electric car's battery, supplementing the energy from an electric receptacle.[5,6] The significance of the drawing is its early date for considering use of power from the sun to drive the wheels of an automobile.

*Figure 12.1. The roof-located solar-array powers accessories only. (Ref. 3)*

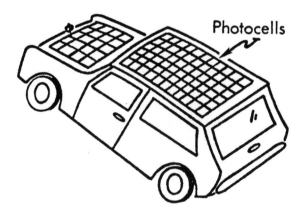

*Figure 12.2. One of the earliest drawings of solar cells for supplemental motive power. (Ref. 2)*

In calculating the assist from solar panels and utilizing Figure 12.2 as a model, allow 4 m² on the rooftop and hood. The vertical sun irradiation is approximately 1 kW/m².[7] Therefore, under the best conditions, the cells receive energy at the rate of 4 kW. Because the efficiency of these cells is 12 to 15%, the electric power flowing from the cells is now approximately 500 watts.[8] Suppose the useful sun (and its angle effect) is six hours per day in Los Angeles. Approximately 3 kWh of energy would be delivered to the battery daily. Using the 1990-announced General Motors battery-only electric *Impact* as a model, which has an energy consumption at 55 mph (89 km/h) calculated to be 113 watt-hours/mile, the range extension would be $3000 \div 113 = 25$ to 30 miles (40 to 48 km).

Figure 12.3 shows a solar-powered car from Crowder College in Neosho, Missouri. This car was the first to cross America, from west to east, in 1984.

# Freeman's British Solar-Electric Vehicle

In 1981, Alan T. Freeman of the United Kingdom reportedly built a three-wheeled vehicle powered only with solar cells. He covered 13 miles (20.9 km).[9] In 1982, Hans Tholstrup and also Larry and Garry Perkins in Australia made a large leap in the application of solar cells to personal transportation.[10] Their vehicle crossed Australia from west to east. Bearing a bathtub-shaped body (shown in Figures 12.4 and 12.5) and constructed of steel tubing, *The Quiet Achiever* is a four-wheeler equipped with bicycle racing tires to reduce road friction. Braking

*Figure 12.3. A solar-powered car from Crowder College in Neosho, Missouri, was the first to cross America, from west to east, in 1984. Art Boyt is shown as the driver.*
( The Joplin Globe)

*Figure 12.4. With vertical sun,* The Quiet Achiever *cruises at 15 mph (25 km/h).*
*(Ref. 10. Photo courtesy of West Australian Newspapers Ltd.)*

is accomplished by conventional cycle equipment. Curb weight of the vehicle is 276 lb (125 kg). The vehicle is powered with 720 solar cells feeding a motor in which the pinion is chain-connected to the sprocket of the rear drive wheels.

In passage from Perth in western Australia on the Indian Ocean to Sydney on the east coast of Australia, a distance of 2566 miles (4129 km), the best run for the 20-day trip was 191 miles (307 km).[11] The trans-Australia crossing was made in the Southern Hemisphere in summer, January 1983; therefore, travel could last as much as 11 hours per day. Typical cruising speed was 15 mph (24 km/h). One of the drivers of *The Quiet Achiever* was Hans Tholstrup, who would organize in 1987 the Trans-Australia Pentax World Solar Challenge Race discussed in Chapter 13.

Coincidentally on 2 November 1982, I wrote a similar proposal for an American cross-country trek using a solar-powered vehicle. I submitted the proposal to solar cell manufacturers for financial support. This request, illustrated as Table 12.1, fell on deaf ears. With the 1992 proposed World's Fair awarded to Chicago, I presented a similar proposal to the Fair Authority through Heliobat Systems Inc.[12] based on a design by Professor William A. Becker, Department of Design, University of Illinois (Chicago), Professor Alan L. Kistler, Mechanical Engineering Department, Northwestern University (Evanston, Illinois), and me. An illustration of this type of solar cell/battery-powered *Heliobat* (Greek term which loosely means *sun*

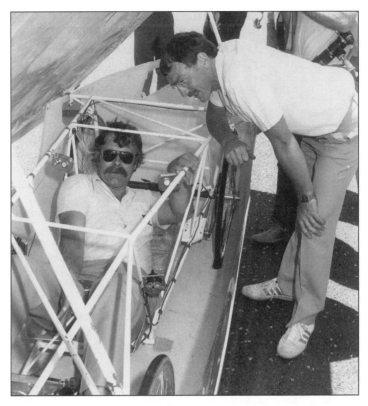

*Figure 12.5. Hans Tholstrup at the wheel. Steel tubing frame supports the fiberglass body. (Ref. 10. Photo courtesy of West Australian Newspapers Ltd.)*

*powered*) is illustrated in Figure 12.6. The proposed vehicle's rooftop would bear 1 m² of solar cells charging a nickel-cadmium battery which in turn powers a separately excited-field d-c electric motor linked to the drive wheel. Used on the Fair Grounds which had a high density of people, speeds are necessarily limited to walking needs. To supplement the charge in the battery, in addition to the flow of current from the solar cells, there would be suitably located in the parking area at each exhibit entrance individual wire coils energized with 400 hertz alternating current. In another locale, this system was developed and reported in detail by Kosrow Lashhkari, Steven E. Shladover, and Edward H. Lechner.[13] A similar coil connected to an onboard battery charger is placed on the underside of the carriage. The use of Faraday's 1821 principle of mutual inductance (with a rectifier) serves to charge the battery, while the person who is renting the *Heliobat* is located in a Fair Pavilion.

207

**TABLE 12.1**
**TENTATIVE SPECIFICATIONS AND BUDGET FOR**
**COAST-TO-COAST PROJECT SOLARWIND**
**2 NOVEMBER 1982**

| | |
|---|---|
| 1. Description of the vehicle | Three-wheel, minimum support solar parallel hybrid vehicle bearing a solar cell canopy, a solar chargeable battery, an electric drive motor, seat, and wind shroud. Weight: 70 to 100 lb. |
| 2. Power source | $5 \ m^2$ of photovoltaic cells to yield 500 peak watts fitted on vehicle canopy. Solar power may flow into the motor or battery. Either may serve to drive the vehicle. |
| 3. Power control | Voltage switching by battery or by panels. |
| 4. Electric motor | Separately excited field. |
| 5. Top speed | Approximately 15 mph (24 km/h). |
| 6. Day's run | Approximately 100 miles (161 km) on average. |
| 7. Cross-country run | One to two months. |
| 8. Anticipated construction, modification, and test time | Twelve months. |
| 9. Tentative cost of the project | $600,000, arrived as follows: |

| | |
|---|---:|
| One test and two backup vehicles @ $20,000 | $ 60,000 |
| Two equivalent man years of high-tech personnel with support @ $150,000 per man year | $300,000 |
| One equivalent man year of high-tech personnel for coast-to-coast dash @ $150,000 per man year with support | $150,000 |
| Sleeping and repair van rental, chase car, and police escort | $ 40,000 |
| Unanticipated | $ 50,000 |

Next, Joel Davidson and Gregory Johanson of Photovoltaic Power Systems designed a solarmobile, *Sunrunner,* which traveled 24.74 mph (39.8 km/h) at Bellflower, California, on 1 July 1984, for a world's speed record.[16] Later, 19 July to 29 August 1984, Chris Kalmbach, Art Boyt, and others from a party of Crowder College, Neosho, Missouri, completed the first trans-America solar car west-to-east crossing from San Diego, California, to Jacksonville, Florida. This vehicle cost $5,000.[17] Another historic solarmobile, a participant in the 1985 *Tour de Sol* race, was entered by Solarex Corporation of Rockville, Maryland, and is illustrated in Figure 12.7.[18] Other vehicles also show the development of solarmobile design.

Next are illustrated and described three world-class solarmobiles at the time, also showing three different approaches for mounting the solar cells: the "fixed-body" type, the "collar-array" style, and the "pool-table" approach. These designs are shown in order, and each has its champions.

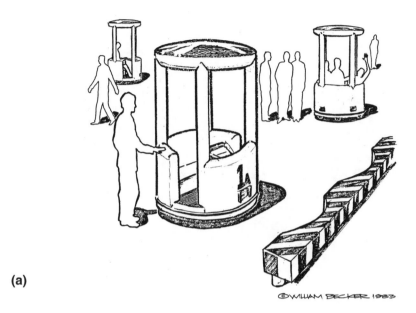

**(a)**

©WILLIAM BECKER 1983

*Figure 12.6. (a) Solar/battery-powered personal vehicles proposed for the anticipated Chicago 1992 World's Fair. Battery charging is accomplished both from rooftop solar cells and by induction from buried coils at pavilions. (William Becker)*

**(b)**

*(b) The Witkar driver-operated electric taxis in Amsterdam were the origin of thinking for the anticipated 1992 Chicago World's Fair. (Refs. 14 and 15. Photo courtesy of David Halperin)*

209

*Figure 12.7.  In the 1985 Swiss Tour de Sol, Solarex placed second.  The car has an array of 285 watts (peak) and is capable of 31 mph (50 km/h).  (Solarex Corporation)*

# The General Motors *Sunraycer*

Reading about the preparation for and the designers chosen for the General Motors *Sunraycer,* any student of electric cars will immediately conclude that the world's largest automobile manufacturer wished to use this vehicle as an instrument to illustrate the competency of its organization.  The plan succeeded admirably.

The *Sunraycer* is a one-seat, four-wheeled car designed especially for the Australian 1987 Transcontinental Race described in Chapter 13.  Using the best talent from Dr. Paul B. MacCready's AeroVironment Inc. in a major role, the GM Hughes Aircraft Company, and other subsidiaries as well as leading technologists from many nations but primarily the United States, GM and its partners in the project created the structure, body, and drive system of the vehicle.  They also covered 90 ft$^2$ (8.0 m$^2$) of the surface of the car with solar cells.

The *Sunraycer*, built on an aluminum tube space-frame, weighs 360 lb (163 kg).  With a driver's weight ballasted to 187 lb (85 kg), the car has a gross weight of 547 lb (248 kg).  The vehicle is 19.7 ft long, 6.6 ft wide, and 3.3 ft high (6 × 2 × 1 m) and is shown in Figure 12.8.

*Figure 12.8.  General Motors 1987* Sunraycer.  *(General Motors Corporation, Detroit)*

The skin of the vehicle is of a lightweight honeycomb-sandwich material, giving rigidity and great strength.  Additional details are provided in Ref. 19.

The electrical output of the solar cell flows to a battery and to a transistor-controlled, three-phase, variable frequency, six-pole, permanent magnet synchronous motor which powers the left rear wheel through a belt drive.  The four wheels are of 20-in. (51-cm) bicycle-type, bearing smooth-surface Korvar tires.  The total vehicle complies with international vehicle regulations for highway driving.[a,20]  The serious student should read both "Lessons of Sunraycer," written by the designers,[21] and *An Analysis of the Components of Power Loss in a Solar-Powered Electric Vehicle.*[22]

With such success, some knowledgeable circles would encourage GM to take elements of this costly vehicle, downgrade others for economy, build a fleet of several dozen, and, under controlled conditions, assign them to a responsible party in a high solar area of one of the fifty states to determine if solar-powered vehicles can indeed be competitively viable, thus attacking the domestic petroleum and air pollution problems, as well as the greenhouse effect.

# The MIT Solarmobile *Solectria IV*

The Massachusetts Institute of Technology (MIT) solar-powered electric vehicle, named *Solectria IV* and shown in Figure 12.9, is a single-seat, three-wheeled race car, the fourth of a series designed by James D. Worden and his associates.  The vehicle was built by students

---

[a] GM built two *Sunraycers*.  The one in the Trans-Australia Race was partially equipped with gallium-arsenide solar cells, which have a greater efficiency than silicon-based cells.  The solar array delivered 1500 watts.

*Figure 12.9. The Massachusetts Institute of Technology Solectria IV.
(Nancy Hazard, 1990 Northeast Energy Association Tour de Sol)*

in the Mechanical Engineering Department shops of MIT in Cambridge, Massachusetts, specifically to participate in the 1987 *Tour de Sol* race in Switzerland. This competition among solarmobiles is believed by MIT authorities to be an excellent training ground for real-life problems in engineering, organization, financing, purchasing, competition, entrepreneurship, and in self-disciplining the participants.[23] Furthermore, with this program, experiences such as the travel involved, the meeting of other contestants from a variety of countries, and the observation of their solutions to similar problems are difficult to teach in the classroom.

*Solectria IV B* was modified from the solarmobile described above specifically for the 1987 Trans-Australian Race instituted by Tholstrup.[10] Among the changes were modifications to the solar array, which was now a tiltable "pool-table" type, and the installation of an a-c permanent magnet synchronous motor and suitable control system. Additional specifications may be found later in this chapter.

# Ford Motor Company of Australia *Sunchaser*

The Ford Motor Company of Australia *et al.* created the solar-powered *Sunchaser* to partici-pate in the Pentax World Solar Challenge Race of 1987. The vehicle is illustrated in Figure 12.10. The design team was led by Jon Retford, whose group had previously produced a lightweight vehicle which set the (then) world's fuel economy record of 5107 mpg (4400 km/l).

Retford and a group of 15 volunteer engineers spent two and one-half years on the *Sunchaser*. Their previous work with lightweight materials, low-rolling-resistance tires, aero-dynamic bod-ies and wheels contributed a sound background for their previous solarmobile project. Among the factors that influenced the design were[24]:

1.  Energy collection capability
2.  Stability
3.  Aerodynamics in various wind conditions
4.  Serviceability
5.  Weight
6.  Reliability
7.  Ease of construction
8.  Budget constraints

In any cross-country race, Retford considered reliability a most important factor; hence, the vehicle was developed on a modular philosophy in having sub-assemblies of all key vehicle systems. These sub-assemblies were: 1) front and rear suspensions packages, 2) gearbox and motor assembly, 3) power electronics, and 4) a photovoltaic collection panel. With the race running from Darwin to Adelaide, or nearly due south, and with the angle of the sun-tracking

*Figure 12.10. Ford Australia 1987* Sunchaser. *(Ford Australia, Ltd.)*

from east to west, a tilting solar array was most desirable to enhance the energy collection ability. The negative effect was somewhat poorer aerodynamics. Also, there was the question of the vehicle remaining stable when encountering large oncoming traffic or in strong side winds. With computer examples of the vehicle and testing of one-quarter-sized models in the wind tunnel at The Royal Melbourne Institute of Technology, the design could be finalized at speeds expected to exceed 100 km/h (62 mph). Further refinement took place on a full-scale vehicle at the Ford Proving Grounds at Lara, 40 miles (64 km) southwest of Melbourne. Among the tests was a satisfactory performance when meeting a semi-trailer at closing speeds of 180 km/h (112 mph).

The Ford *Sunchaser* is a one-seat, four-wheel vehicle of monocoque construction made from Kevlar and carbon fiber materials. The efficiency of the *Sunchaser* array is 15.4%.[25]

Critical system elements of solarmobiles are discussed next.

## Electrical and Mechanical Features of a Solarmobile

The preceding pages described the array characteristics of three racing solarmobiles. However, what is inside the external shell? Figure 12.11 shows the space frame of the 1990 University of Michigan *Sunrunner* and illustrates an operator in driving location. Safety straps are not shown. The steering wheel is before the driver. More distal from his eyes would

*Figure 12.11. Space frame of the 1990 University of Michigan* Sunrunner.
*(Mary Longbrake)*

be the instrument panel, and, finally, engaging his feet are the brake and accelerating pedals. Figure 12.12 demonstrates the intricacies of the solar cells on *Sunrunner*, the largest single material cost of a solarmobile if silver-based batteries are avoided. One of the main features of a solarmobile is that it must be lightweight. Note the exposed framing, and also observe in Figure 12.13 the enclosed wheel spokes to enhance aerodynamics on the 1987 GM *Sunraycer.*

Figure 12.14 more clearly illustrates the internal framing of a solarmobile. In that photo, students in the Mechanical Engineering Department of Massachusetts Institute of Technology (MIT) in Cambridge, Massachusetts, are listening to Professor Harry West explain a structural problem. In Figure 12.15, James D. Worden, then the student designer, contemplates an electrical difficulty. Notice the battery complement is directly behind the protecting roll-bar

*Figure 12.12. Solar cells on 1990 University of Michigan* Sunrunner.
*(Mary Longbrake)*

215

*Figure 12.13. Enclosed wheels of* Sunraycer. *(Bruce McCristal)*

*Figure 12.14. Steel framing of 1987 MIT* Solectria IV. *(Harry West)*

against which the driver's head and shoulders are supported. Finally, Figure 12.16 illustrates two of the three compound-wound d-c motors chain-linked to the transmission. (Currently, solarmobile designers are moving toward employing a single a-c drive motor.) Cryptic rust in the transmission down-ranked Worden's finishing position in the 1987 *Tour de Sol.*

## American Solar-Assist Electric Motorcars

Finally, in Chapter 14, Figure 14.2 shows in the background two St. Johnsbury Academy student-built and a Solectria Corporation *solar-assist* electric cars, indicating the cutting edge of *solar-assist* and *solar-dominant* automobiles is from academe or small business organizations. St. Johnsbury Academy is located in Vermont, and its vehicle, *The Electric Hilltoppers* and *The Electric Jewel* received first and second place prizes from the U.S. Department of Energy at the 1992 American *Tour de Sol* for the best student-built cars. The Solectria Corporation *Force GT* of Arlington, Massachusetts, also pictured in Figure 14.2, used nickel-cadmium batteries. It averaged 75 miles (121 km) per day and took both the Most Efficient and the Range Prizes by driving slightly more than 100 miles (161 km) on a single battery

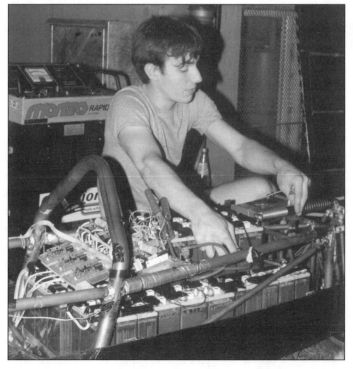

*Figure 12.15. The battery complement and its nesting in* Solectria IV. *(Harry West)*

*Figure 12.16. The d-c motor drive system of* Solectria IV. *(Harry West)*

charge, using 147 watt-hours per mile while carrying three people and traveling at 35 mph (56.3 km/h), as reported by Nancy Hazard of Northeast Energy Association of Greenfield, Massachusetts.

# 1996 Solar-Battery Ford *Festiva,* Designed for Saminco Inc.

We might ask, "How helpful is the solar component in solar-assist vehicles?" Bonne W. Posma, President of Saminco Inc. in Fort Meyers, Florida, forwarded pictures of his solar-assist Ford *Festiva* (Figure 12.17). The car appears to bear 2.5 m$^2$ of silicon solar cells. Assuming a vertical sun of 1000 watts/m$^2$, the cells of his car received 2500 watts of sunshine. Assuming a cell efficiency of 15%, 375 watts-electric theoretically reach the battery. Referring to Figure 11.2, Fort Meyers receives an average of 4.7 peak sun-hours per day. Delivered to the battery would be 4.7 × 375 = 1760 watt-hours. Assume in city driving the electric *Festiva* requires 200 watt-hours per mile. Therefore, the solar component has a value of roughly 8 miles (13 km) per day in Fort Meyers.

# Japanese Solar-Assist Automobiles

Solar-electric automobiles are divided into two categories, as previously delineated: 1) solar-assist vehicles, and 2) solar-dominant vehicles. Both are receiving emphasis in shops in the industrial world to fill special niche rolls. Japanese companies have been particularly active in developing solar-assist cars.

*Figure 12.17. A solar-assist, battery-powered Ford* Festiva, *1996. (Bonne W. Posma, Saminco Inc.)*

Tohuko Electric Power Company's *MYLD* (for *My Lovely Drive*) is a compact two-seater solar-assist car fueled by a 20-cell nickel-iron battery. A roof-mounted 90-watt solar array contributes to the battery charge, providing perhaps 600 to 800 watt-hours per day.[26] This amount of energy is expected to increase the range of the vehicle only 2 to 4 miles (3 to 6 km) per day.

The *Mirai-1,* developed by the Sanyo Electric Company (and described in Chapter 5), is a solar-assist, 880-lb (360-kg), two-seat electric motorcar securing power from "an environmentally safe" chemical fuel cell supplemented with nickel-cadmium batteries.[27] The fuel cell supplies the long-term energy; the batteries provide accelerating power. An amorphous silicon solar cell array on the hood charges the batteries during daylight. The company spent a reported 50 million yen ($500,000) on development of the vehicle.

The next chapter discusses the *Australian World Solar Challenge* of 1987, a triennial race that was the first to introduce true professionalism into solar car racing.

# References

1. Wakefield, Ernest H., *History of the Electric Automobile: Battery-Only Powered Cars*, Society of Automotive Engineers, Warrendale, PA, 1994, pp. 149–189.
2. Wakefield, Ernest H., *The Consumer's Electric Car*, Ann Arbor Science Publishers, Ann Arbor, MI, 1977.
3. "First Application for Solar Power Charge," *Electric Vehicles*, **60,** 1 March 1974, pp. 8–9.
4. Wakefield, Ernest H., *The Consumer's Electric Car*, Ann Arbor Science Publishers, Ann Arbor, MI, 1977, pp. 114–115.
5. Fujinaka, Masaharu, "Future Vehicles Will Run with Solar Energy," SAE Paper No. 891661, Conference on Future Transportation Technology, Vancouver, British Columbia, Canada, Society of Automotive Engineers, Warrendale, PA, 1989.
6. Kumagi, Naotake, and Tatemoto, Minoru, "Application of Solar Cells to the Automobile," SAE Paper No. 891696, Conference on Future Transportation Technology, Vancouver, British Columbia, Canada, Society of Automotive Engineers, Warrendale, PA, 1989.
7. Chalmers, Bruce, "The Photovoltaic Generation of Electricity," *Scientific American,* **235,** 4 October 1976.
8. Franck, Robert I., and Kaplow, Roy, "Performance of a New High-Intensity Silicon Solar Cell," *Applied Phys. Lett.*, **34**, 11 January 1979.
9. *Electric Vehicle News,* **10**, 4 November 1981, p. 32.
10. Tholstrup, Hans, and Perkins, Larry, "Across Australia by Sun Power," *National Geographic,* **164,** 5 November 1983, pp. 600–607.
11. *Science,* **83**, July/August 1983, p. 10.
12. Letter to Thomas Ayers, Chairman, 1992 Chicago World's Fair Committee, from Dr. Ernest H. Wakefield, Chairman, Heliobat Systems, Inc., 12 July 1983.
13. Lashhkari, Kosrow, Shladover, Steven E., and Lechner, Edward H., "Inductive Power Transfer to an Electric Vehicle," Eighth International Electric Vehicle Symposium, Washington, DC, 1986.
14. Senger, W.M., "Silent Car in Amsterdam," *Environment*, Vol. 16, No. 8, October 1974, pp. 14–17.
15. Wakefield, Ernest H., *History of the Electric Automobile: Battery-Only Powered Cars*, Society of Automotive Engineers, Warrendale, PA, 1994, p. 365.
16. *Guinness Book of World Records*, Sterling Publishing, NY, 1989, p. 190.
17. Boyt, Arthur, personal communication, 14 November 1988.
18. Goldsmith, John V., Solarex Corporation, personal communication, 15 June 1988.
19. MacCready, Paul, *GM Sunraycer Case History,* Society of Automotive Engineers, Warrendale, PA, 1988.
20. Brooks, Alec N., personal communication, 25 March 1988.
21. Wilson, Howard G., MacCready, Paul B., and Kyle, Chester R., "Lessons of Sunraycer," *Scientific American,* March 1989, pp. 90–97.

22. Patterson, D.J., *An Analysis of the Components of Power Loss in a Solar-Powered Electric Vehicle,* Darwin Institute of Technology, Darwin, Australia, 30 January 1988.
23. West, Harry, "*Tour de Sol* '87: A Report on the MIT Entry in the 1987 *Tour de Sol,*" Massachusetts Institute of Technology, Cambridge, MA, November 1987.
24. Retford, Jon D., Ford Motor Company of Australia, personal communication, 24 May 1988.
25. Johnson, Kalwey A., Solarex Corporation, personal communication, 3 November 1988.
26. *Popular Science*, February 1992, p. 52.
27. *The New York Times*, 19 March 1992, p. C5.

# CHAPTER 13

# The Trans-Australian World Solar Challenge Race—1987

## Background

The 1985 Swiss *Tour de Sol* first introduced racing to solar-powered cars. However, it was the 1987 Australian World Solar Challenge Race that injected true professionalism into the design of vehicles for this class of competitive racing. (See Ref. 1 for additional information on the race.) To trace this transition, Dennis Bartell had recently completed his epic 1900-mile (3058-km), four-month walking trek across the Australian Outback. His journey stretched from Darwin on the Gulf of Carpentaria in the north to Adelaide on the Gulf of St. Vincent in the south, in celebration of South Australia's Jubilee 150, as more fully described in Chapter 9 and shown in Figure 9.1. With all the sunshine experienced while crossing this largely desert land, Bartell wondered how this solar energy might be effectively used.

On arrival at his Adelaide home, Bartell convinced financial backers to have Terry Trevor design and the Natural Energy Company (a division of Pecan Engineering in suburban Brampton) build a $40,000 solar-powered electric car, the *Spirit of Adelaide*. With this vehicle, Bartell would make a north-to-south crossing from Darwin to Adelaide on the soon to be completed Stuart Highway that links the two cities by way of centrally located Alice Springs. In one stroke, reverence would be paid John McDouall Stuart who, as a pioneer with others, first made this overland trek by foot and horse. At the same time, it would celebrate the centennial of the first gasoline-powered car in Australia. As a precedent, the continent had been crossed from west to east by *The Quiet Achiever* in 1983, as cited in Chapter 12. Bartell's resulting *Spirit of Adelaide* was a one-seat, four-wheeled vehicle, 5 m (16 ft) long, and 1.8 m (5.9 ft) wide, bearing a solar array capable of delivering 810 watts with a zenith sun. The vehicle's maximum speed was 48 mph (77 km/h) with the car powered by four batteries, charged by the array, driving a 7.5 hp (5.6 kW) motor. A fifth battery operated the accessory equipment. The solarmobile was delivered to Bartell. His plan was to leave Darwin on 11 November 1986 and follow the Stuart Highway to Adelaide by way of Alice Springs, arriving home around 5 December 1986. Figure 13.1 shows Bartell and his car.

True to his promise, Bartell left Darwin on 11 November 1986. After the usual problems associated with insufficiently tested mechanical and electrical items, together with unseasonable weather, Bartell arrived to a champagne city-welcome in Adelaide on 20 December 1986, thus establishing a first solar-powered vehicle crossing of Australia from north to south. This accomplishment extended the remarkable record of Australian battery-only electric car development as recounted in Chapter 22 of *History of the Electric Vehicle: Battery-Only Powered Cars*.[2] Bartell's *tour de force* spurred more interest. With this proven financial interest and impetus, Hans Tholstrup, who was instrumental in the earlier west-to-east solar-powered vehicle crossing, had observed the media success of the *Tour de Sol* race in Switzerland in 1985. He promoted and received backing for what, after some difficulties, would eventually become the Pentax World Solar Challenge Race of 1987. The event became a joint Australian–Japanese promotion sponsored by the Japanese firm, Asahi Optical Ltd.[3] The Broken Hill Associated Smelters, whose main operation is closely passed by the Stuart Highway, offered a $25,000 perpetual, gold and silver trophy that was 3 ft (1 m) tall. The race, suggested for every third year, yields no prize money. With Tholstrup and others promoting it, the World Solar Challenge Race became an ever-growing "green" crusade which was to attract twenty-four entries from seven nations, including two high-school-constructed vehicles from Australia. Solar-powered cars were entered from Australia, Denmark, Japan, Pakistan, Switzerland, West Germany, and the United States.

Meanwhile, relying heavily on the more experienced group running the Swiss *Tour de Sol*, Tholstrup had established vehicle specifications for the 1987 Pentax World Solar Challenge Race: the area of solar array could be no greater than 8 m² (86 ft²), the length of vehicle was limited to 6 m (20 ft), the width was limited to 2 m (6.5 ft), and the height could be no greater than 2 m (6.5 ft). Only solar power would be permitted both in charging the batteries and in vehicle operation. As in the Swiss race, all national traffic rules were to be followed, the vehicles were to be equipped with published stated accessories, and the vehicle drivers were to be ballasted to 187 lb (85 kg).

As the media readied for the race, *The Advertiser* (Adelaide) on 10 July 1987 pictured the Ford Australia *Sunchaser* car entry (see Figure 12.10 in Chapter 12).[4] On 7 August, *The Advertiser* showed the General Motors *Sunraycer* (Figure 12.8 in Chapter 12).

*Figure 13.1. Bartell (top) and his car (bottom) triggered the trans-Australian solar-electric car race from Darwin to Adelaide over the Stuart Highway.* (The Advertiser, *Adelaide, Australia)*

With race time nearing, the 29 October edition of *The Darwin Northern Territorian* warned high-altitude, northern-hemispheric drivers, whose cars bore streamline bubble cabins, that the noonday shade-temperature south of Darwin can reach 104°F (40°C).[5] A later issue of the newspaper reported that on the road to Darwin, the solar-electric car *Alarus* (Figure 13.2), built by Frank Castino and Dimitri Ladovic of Strathfield, Australia, was smashed when its transportation truck was overturned. (*Alarus* was rebuilt.) *The Advertiser* of 30 October quotes Dr. Paul B. MacCready, GM *Sunraycer's* Project Manager introduced in Chapter 12, as saying:

> A unique race like this is a symbol, a catalyst. It's like Lindberg's flight over the Atlantic. That (trip) didn't break any new technology or provide new planes. What it did do, though, was to change the way people thought. That's more important.

Of historical interest, note that E.P. Ingersoll, editor and proprietor of *The Horseless Age*, the first magazine devoted to motor vehicles that were soon to be commonly called *automobiles,* wrote a similar paragraph after the Paris-Bordeaux-Paris automobile race of 1895.[6]

On the day before the contest began, the 31 October edition of *The Advertiser* offered warnings again to the overseas racers about the 104°F (40°C) heat, the possible effect on lightweight cars of the "wind wash of 130-tonne road-trains" with three trailers traveling at 75 mph (120 km/h), and the "presence of wandering buffaloes" because the highway was not fenced.

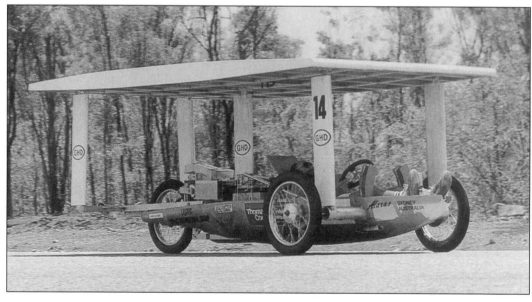

*Figure 13.2. The* Alarus. *(Bill Tuckey)*

The Pentax World Solar Challenge Race of 1 November 1987 was to match, in a sense, the *Tour de Sol* competition in Switzerland, but in a dry, hot climate of lesser altitude, with far greater course length. In Australia, which is in the southern hemisphere, the race date corresponds to 1 May in North America or in Europe, a period of long daylight. The starting point for the race was the city of Darwin. Termination of the race was near Adelaide, a distance of approximately 1980 miles (3200 km), "down the track" as the Australians say, along the newly completed all-weather macadam Stuart Highway. In truth, the race was a celebration of the bicentennial founding of this southern Commonwealth. More than half the race course has an annual rainfall less than 10 inches (25 cm), and the remainder has little more. This fact promised a reasonable chance for sunny skies for the solar-powered vehicles. In the competition, entrants raced from 8 A.M. to 5 P.M. every day. There were three classes of entrants: 1) commercial, 2) schools and institutions, and 3) individuals. With the exception of these criteria, the only limit was that all power for the vehicles must be solar originated. Hence, the battery must be solar-charged. To ensure compliance, a race committeeperson accompanied each entry's service car.

# The Race Begins

Twenty-three entries from the seven countries cited were on the starting line in Darwin's Casuarina Shopping Square at 9 A.M. Sunday. Waiting in the shadows were the vans carrying the race-team support groups: spare driver, repair engineers, a race monitor, extra parts, and desirable accessories. Ahead of the race teams loomed searing Australian Outback heat, driver fatigue, electrical and mechanical failure, wandering cattle, and a plague of flies. For one challenger, a collision would occur with an automobile in Alice Springs, an event which may have been the first collision in history between solar-electric and gasoline-powered cars. Many of the drivers were veterans from other solar car races. For example, James D. Worden, a junior (in 1987) at the Massachusetts Institute of Technology (MIT), had led a design and construction team which had built four solar-powered racing cars. The current vehicle was a modification of the *Solectria IV,* which had four months earlier competed in the *Tour de Sol* in Switzerland. Likewise, the Swiss entry, *Spirit of Biel* (Figure 13.3), with "pool-table" array, had placed fourth in the 250-mile (400-km) Swiss race a few months earlier and had won the previous year.

The cost of building the vehicles varied from $5,000 for *Solar Samba* (Figure 13.4), a car built in eighteen months (according to Syed Attique Shafaat, the group's leader) at the University of Engineering and Technology in Karachi, Pakistan, to an estimated $3 million expended on the General Motors *Sunraycer,* which was constructed and tested over a seven-month period. The latter vehicle had established on 17 September 1987 in Arizona a world record speed for solar-electric cars of 34.2 mph (55 km/h) on an overcast day. In a test to determine the starting order of the race, the *Sunraycer,* in the bright sun of Darwin, had reached a speed of 70 mph (113 km/h) and had earned the pole position. John Paul Michell's *Mana La*

227

*Figure 13.3.* Spirit of Biel, *with "pool-table" array.*
*(Bruce McCristal)*

*Figure 13.4.* Solar Samba, *University of Engineering and Technology, Karachi, Pakistan. (Bruce McCristal)*

(Figure 13.5), was second at 57.8 mph (93 km/h). Other competitors clocked lower speeds with radar; indeed, *Southern Cross* (Figure 13.6) and *Just Magic* (Figure 13.7) went so slowly that they did not register on the radar dial.

For the 2 November issue of *The Darwin Northern Territorian,* a reporter wrote:

> Darwin turned out a crowd of 10,000 and a canopy of powerful sunshine as the 23 high-tech competitors in the Pentax World Solar Car Challenge headed south yesterday. The fleet includes some of the rarest road models ever to invade the motoring world. Some, including the beetle-like GM entry, *Sunraycer,* made an immediate impact speeding past lines of onlookers immediately after the 9 A.M. start...*Sunraycer* is widely tipped to win the challenge (in its class), being run over the lonely 3080 km (1914 mile)(course) between Darwin and Adelaide...

On the next day, the same reporter related the following:

> The sleek GM *Sunraycer* continued its powerful run down the Stuart Highway today, well ahead of its opponents in the inaugural Pentax World Solar Car Challenge. *Sunraycer* surged through tiny Elliott at 11 A.M. on its way to its next major destination of Tennant

*Figure 13.5.* Mana La, *John Paul Mitchell, Hawaii. (Bruce McCristal)*

*Figure 13.6.* Southern Cross, *Semiconductor Energy Laboratory, Japan. (Bruce McCristal)*

*Figure 13.7.* Just Magic, *Goodwood High School, Adelaide, Australia. (Bruce McCristal)*

Creek. Switzerland's *Spirit of Biel* has edged from third into second place in front of the Ford entry. And late this morning, Hawaii's *Mana La* was in fourth place, followed by Australian Geographic's *Marsupial* (Figure 13.8). The Darwin Institute of Technology *Desert Rose* (Figure 13.9) is proving a strong challenger in seventh place behind the West German *Lichtblick 2* (Figure 13.10).

On this leg of the race, the Stuart Highway rises slowly as it climbs the starkly dry Barkly Tableland, heading for central Australia.

The 4 November issue of *The Darwin Northern Territorian* reported progress on the third day of the race:

> *Sunraycer* had reached Barrow Creek (elevation: 300 m, or approximately 1000 ft) south of Tennant Creek by late yesterday and is expected to be in Alice Springs this afternoon. Race watchers believe she can reach Adelaide well before Sunday (the seventh day).

South from Barrow Creek, the highway climbs the MacDonnel Range, which has peaks near Alice Springs crowning to almost 5000 ft (1500 m). However, the town has a considerably lower elevation. The report continues:

> The Swiss entry, *Spirit of Biel*, and the Ford Australia car are maintaining second and third positions, respectively...The Hawaiian, *Mana La*, in fourth place early yesterday, has been plagued by battery problems and is back in seventh spot. The Territory entry, the Darwin Institute of Technology's *Desert Rose*, has been tussling with the Dick Smith-driven

*Figure 13.8.* Marsupial, *Australian Geographic. (Bruce McCristal)*

*Figure 13.9.* Desert Rose, *Darwin Institute of Technology, Darwin, Australia. (Bruce McCristal)*

*Figure 13.10.* Lichtblick 2, *Rolf Disch, Germany. (Bruce McCristal)*

*Marsupial* for fourth place. The *Rose* held (position four) late yesterday, but the *Marsupial* has made a run and is 27 km (16 miles) ahead of the Darwin entry. (Other) challengers are reported to be well back.

On 5 November, *The Darwin Northern Territorian* reported an unusual sight, also captured in Figure 13.11:

South of the (Northern Territory—South Australia) border (and edging on the Simpson Desert), the sleek GM *Sunraycer*, blitzing the field...passed the NT (Northern Territory)—South Australia police camel expedition. The modern-day cameleers are doing it (the Darwin—Adelaide trek) in olden-times style for (Australian) Bicentenary purposes. They are not expected to reach Adelaide until January (two months later). *Sunraycer* should be there by the weekend.

The same day, the paper reported several mishaps:

The Swiss entry...crashed in Alice Springs this morning. The *Spirit of Biel* and a conventional automobile collided in the middle of town about 8:45 A.M. The...(former) received substantial damage. It was the second mishap today among vehicles competing in the 3000-km (1864-mile) solar-car race from Darwin to Adelaide. Another overseas entry was reported damaged after overturning near Tennant Creek. The vehicle is believed to be the Crowder College (U.S.A.) entry, *Star* (Figure 13.12). (In describing the car accident,) The *Spirit of Biel* was turning right at a main intersection in Alice Springs when the collision occurred. The Swiss car was in second place before the accident, about 70 km (44 miles) ahead of the third-place Ford entry. Ford has moved into second place, but the Swiss team was hoping to have its car back on the road about an hour after Ford's *Sunchaser* left Alice Springs.

*Figure 13.11. The GM* Sunraycer *passes a camel train near Alice Springs. (Bruce McCristal)*

*Figure 13.12. Star, Crowder College, Neosho, Missouri, the first day of the race.
(Arthur Boyt, Crowder College)*

The race was won by the *Sunraycer* on Friday, 6 November. Several other entries were about two days behind. The 6 November issue of *The Darwin Northern Territorian* reported on their progress:

> ...All but the leading car (GM *Sunraycer*) were hit by rain storms with some hail just south of Tennant Creek and have been held up at water crossings and by lack of sun. (Figure 13.13 shows Ford *Sunchaser* in road water.) At the time of going to press, *Marsupial* was at Barrow Creek...The unstoppable GM entry, *Sunraycer*, reached Adelaide today, blitzing the field...by at least two days. At the time of going to press, *Sunraycer* was heading toward the finishing line in the Barossa Valley's town of Seppeltsfield, about 60 km (36 miles) north of Adelaide. It is expected to be Sunday before the next cars finish.
>
> The ill-fated Swiss driver, Mr. Ernst Fuhrer, has attracted the wrath of Alice Springs police and could be charged with ignoring a "give way" sign. Mr. Fuhrer was injured when the Swiss entry, *Spirit of Biel,* and an automobile collided in Alice Springs yesterday... Mr. Fuhrer, his team, and car left Alice Springs yesterday afternoon to regain lost ground after a rush to repair damage. The *Spirit of Biel* is attempting to reclaim second place from Ford's *Sunchaser,* which moved up from third position after the Swiss mishap. Australian Geographic's *Marsupial* left Alice Springs this morning. The popular Darwin Institute of Technology's entry, *Desert Rose*, was due there about 3 P.M. and is comfortably in fifth place.

*Figure 13.13. Ford Australia* Sunchaser, *splashing road water from which only the GM* Sunraycer *escaped. The storm was a major factor in the race. (Bill Tuckey)*

After the *Sunraycer* slid by the North Flinders Range of mountains in Southern Australia and approached Gulf of St. Vincent, the shore of which faces Adelaide on the west, a reporter would write in the 7 November edition of *The Darwin Northern Territorian*:

> They called it (*Sunraycer*) the *Cockroach*, but what is probably the most technically advanced car in Australia drove into Adelaide yesterday, remarkably rapidly and with an eerie lack of noise. School groups, photographers, and clusters of people lined the routes along Port Wakefield Road and through the city to watch the General Motors *Sunraycer* pass through on its way to the finishing line of the Pentax World Solar Challenge Race for solar cars from Darwin to Adelaide.

# Entries and Results of the 1987 Australian World Solar Challenge Race

Table 13.1 is the entry list of twenty-four solarmobiles from seven countries.[7] Table 13.2 lists the placement of the vehicles that finished (including one car—*Vapor 1*—which did not start).[8]

## TABLE 13.1
## 1987 PENTAX WORLD SOLAR CHALLENGE RACE ENTRIES

| | | Organization | Manager |
|---|---|---|---|
| **Australia** | | | |
| 1. | *Desert Cat* | Chisholm Institute of Technology | Paul Wellington |
| 2. | *Model Sunchaser* | Ford Motor Company of Australia | Jon Retford |
| 3. | *Photon Flyer* | Morphett Vale High School, Adelaide | David Milne |
| 4. | *Alarus* (aboriginal— Daylight) | Frank Castino and Dimitri Ladovic | N/A |
| 5. | *Marsupial* | Australian Geographic | Graham Allen |
| 6. | No Name | Warragul Technical School | Ted Mellor |
| 7. | *Vapor 1* | Clisby Solar Steam Team | Harold Clisby |
| 8. | No Name | Solar Resource Syndicate | Ian Landon Smith |
| 9. | *Desert Rose* | Darwin Institute of Technology | Dr. Jim Patterson |
| 10. | *Just Magic* | Goodwood High School, Adelaide | Tom van Ruth |
| **Denmark** | | | |
| 11. | *Chariot of the Sun* | Sonderborg Teknikum | John Tulloch |
| **Japan** | | | |
| 12. | *Phoebus II* | Hoxan Corporation (Manufacturer of solar panels) | N/A |
| 13. | *Solar Japan* | Nippon TV, Leyton House, Mitsubishi | N/A |
| 14. | *Southern Cross* | SEL—Semiconductor Energy Laboratory | Shinji Imato |
| 15. | *Hama Zero* | Hama (group of inventors) | N/A |
| **Pakistan** | | | |
| 16. | *Solar Samba* | Syed Attique Shafaat, University of Engineering and Technology in Karachi | N/A |
| **Switzerland** | | | |
| 17. | *Spirit of Biel* | Ingenieurschule Biel | Fredy Sidler |
| **U.S.A.** | | | |
| 18. | *Sunraycer* | General Motors Corporation | Ray Borrett |
| 19. | *Mana La* (*Power of the Sun*) | John Paul Mitchell Systems | John Worcester |
| 20. | *Star* | Crowder College | Art Boyt |
| 21. | *Solectria IV-B* | Massachusetts Institute of Technology | James Worden |
| **West Germany** | | | |
| 22. | *Lichtblick 2* | Rolf Disch (Individual) | Rolf Disch |
| 23. | *HelioDet* | Detlef Schmitz | N/A |
| 24. | *Silver Wing* | Michael Trykowski | Michael Trykowski |

Three Entry Categories: 1) Commercial; 2) Institutions—Universities, Schools; and 3) Private—Individuals.

**TABLE 13.2**
**RESULTS OF THE 1987 WORLD SOLAR CHALLENGE RACE**

| Place | Name | Owner | Nation | Time (hrs) (min) | | Avg. Speed (km/h) |
|-------|------|-------|--------|------|------|------|
| 1. | *Sunraycer* | General Motors Corporation | USA | 44. | 54. | 66.92 |
| 2. | *Sunchaser* | Ford Australia | Australia | 67. | 32. | 44.63 |
| 3. | *Spirit of Biel* | Ingenieurschule Biel | Switzerland | 69. | 58. | 42.94 |
| 4. | *Marsupial* | Australian Geographic | Australia | 81. | 26. | 36.90 |
| 5. | *Desert Rose* | Darwin Institute of Technology | Australia | 95. | 27. | 31.48 |
| 6. | *Desert Cat* | Chisholm Institute of Technology | Australia | 98. | 12. | 30.60 |
| 7. | No Name | Solar Resource Syndicate | Australia | 117. | 05. | 25.64 |
| 8. | *Star* | Crowder College | USA | Distance*2424. km | | |
| 9. | *Solectria IV-B* | Massachusetts Institute of Technology | USA | Distance*2399. km | | |
| 10. | *Alarus* | Frank Castino and Dimitri Ladovic | Australia | 146. | 27. | 20.51 |
| 11. | *Chariot of the Sun* | Solvogn | Denmark | 150. | 35. | 19.95 |
| 12. | *Phoebus II* | Hoxan Corporation | Japan | 153. | 31. | 19.57 |
| 13. | *Photon Flyer* | Morphett Vale High School | Australia | 189. | 04. | 15.89 |
| 14. | *Southern Cross* | Semiconductor Energy Laboratory | Japan | 279. | 21. | 10.75 |

* Places with only distance are placings on the fifth day after first car finished. Other placings are solarmobiles which traveled the total distance of 3004 km (1867 miles) after 11 November.

The 7 November story in *The Darwin Northern Territorian* continues:

> Incredibly, the *Sunraycer* finished the 3000 km (1864 miles) (race) about 1000 km (620 miles) ahead of its nearest rivals, the Ford *Sunchaser* and the Swiss *Spirit of Biel,* which are fighting for second place (2900 km—1800 miles—ahead of the Japanese entry *Southern Cross*). (The GM entry) had time to do a victory lap of the city (Adelaide) and stop for a media conference before heading for the finish at Seppeltsfield.

# Comments from Winning Team Members

The GM *Sunraycer* crossed the finish line near Adelaide at 10:04 A.M. on Friday, 6 November, having started in Darwin at 9 A.M. the previous Sunday, a five-day trip. A delighted Mr. Robert Stempel, President of General Motors, was quoted in the 7 November edition of *The Advertiser* (Adelaide) as saying:

> It's a victory for technology and people. I think solar cars for general use are a long way in the future, but really, the idea is to show what you can do with technology, which is to drive a car across this country, powered by God's energy.

The driver, Mr. John Harvey, told the same newspaper when the car paused among the huge crowds gathered at the picturesque Adelaide Oval:

> It was a great welcome for more than 160 km (96 miles) out of Adelaide.  The school kids on the way appeared to be sensing we were making history here; the very idea...we had traveled the 3000 km (1864 miles) from Darwin on solar power is pretty remarkable.  For me, it was very emotional; it was something I will never forget.

*Sunraycer* had covered 3004 km (1875 miles) at an average speed of 42 mph (70 km/h), with no problems except three flat tires.  Harvey concluded:

> This was one of the most significant motoring events in history, and we (GM) had to be here.

Harvey, an Australian touring car driver, had spent two months testing the *Sunraycer* at GM's American desert proving ground in Arizona before taking part in the race as the lead GM driver.  As reported in *The Darwin Northern Territorian* 10 November 1987 after the conclusion of the race:

> *Sunraycer's* technology could also be applied to normal petrol-driven cars.

In commenting on the race, Hans Tholstrup stated, in part:

> It was a great race.  The vehicles were innovative, the crews magnificent, and the race proved solar power has a future in transportation.

The world also was interested in the competition, a fact "indicated by the presence of four television crews from Japan, two from Germany, one from Switzerland, and one from the U.S.A.  And it should not go unnoticed that the *Photon Flyer* (Figure 13.14) was the first solar-powered car built by high-school pupils to cross Australia."

Mr. Tholstrup indicated the next trans-Australian race would occur in the year 1990.  An analysis of the 1990 race follows.

# Performance of Cars

For as grueling a race as was the Trans-Australian World Solar Challenge Race of 1987, we are fortunate to have the analysis of Peter O. Fries of St. Lucia, Queensland, Australia, printed here (with permission) from "A Technical Report for the New Energy Development Organization of Japan."[8]  For the Swiss 1987 *Tour de Sol,* the reader should refer to the report (omitted here for lack of space) of Professor Harry West of Massachusetts Institute of Technology.  Fries writes about the 1987 race, in part[8]:

> As could be expected, the performance of such a wide range of designs varied widely— and in a sense, the race was an ideal solar laboratory 3000 km (1900 miles) long.  Consistent with most scientific endeavors, more questions were asked than answered.  Teams had to

*Figure 13.14.* Photon Flyer, *Morphett Vale High School, Australia. (Bruce McCristal)*

consider hundreds of factors to successfully build a car that would just finish the race—let alone surpass all the other (entrants). Perhaps the key performance factor was reliability. This (factor) demanded the use of simple and robust electronic and mechanical components that could operate as intended and be changed quickly and easily replaced in the event of a malfunction.

Several cars would have finished much higher in the rankings with the proper attention to these criteria. The *Mana La* from Hawaii is a case in point whose battery mismanagement damaged their silver-zinc batteries on the first day and made them so slow that they were forced to retire from the race. Another fast design, the *Solectria IV-B* from MIT, could have finished higher with more reliable electronics.

A summary of the finishing places of the vehicles is given (in Table 13.2). For the purposes of this report, additional analysis is given of the first six finishers and the last place finisher. Because the top six finishing cars were reasonably reliable, their relative finish was determined by panel (solar array) design, power electronics, aerodynamics, weight control, and rolling friction.

### General Motors

The superiority of the GM *Sunraycer* became obvious at the start of the race. The vehicle had the technical advantages of higher efficiency cells, efficient power electronics, and the Magnaquench (permanent magnet synchronous) motor that could be overloaded (because of cooling fins) for extended periods without significant overheating. GM made an opening dash that caught a sun wave which they surfed all the way to the finish line at Gepps Cross. The GM power system delivered over 900 watts average mechanical power to the rear drive wheel, which was double that of the next finisher (Ford Australia.)

A simple math model involving aerodynamic drag area, weight, and rolling resistance was used by Dr. (Chester R.) Kyle to analyze the performance of the top six racers.[9] His analysis shows that the superior performance of the GM car was primarily due to the increased output from the solar array. This explanation was determined by changing each design factor by 10% and calculating the net result from the power equations. (The summary is given in Table 13.3.)

## TABLE 13.3
## GENERAL MOTORS *SUNRAYCER*
## EFFECT OF CHANGING DESIGN*

| | |
|---|---|
| Predicted speed using 838 watts, no change | 66.92 km/h |
| Speed with 10% more power, 921.8 watts | 69.80 |
| Speed with 10% less wind resistance, 838 watts | 68.86 |
| Speed with 10% less weight, 838 watts | 68.09 |
| Speed with 10% less rolling resistance, 838 watts | 68.09 |

*As recalculated by Dr. Chester R. Kyle (personal communication, 1 December 1988).

The calculations are taken for the race distance of 3004 km (1867 miles) and show that the net mechanical power is the most important design factor, followed by wind resistance, weight, and rolling resistance. Table 13.4 shows the predicted performance of the cars at an average power of the second place finisher, Ford Australia. However, there are some interesting aspects on the performance of arrays that are fixed or tilted.

A tilting panel can intercept more direct solar radiation than a fixed panel during running hours. However, before and after running hours, arrays would be equal as both can be adjusted to face the sun. The net result is, the GM arched array would intercept 10% less solar radiation than the tilting array and 3% more that a purely flat array. However, as solar radiation conditions become more diffuse, the advantages between arrays diminish.

Dr. Kyle's analysis shows that if the GM car had used the same type of solar array and electrical equipment as the other cars, the actual design concept would not have been so clearly superior with a much closer race among the top three cars. Even with lead-acid batteries replacing the silver-zinc batteries, the results would have been essentially the same. The GM team did, however, exhibit a superior integrated approach to the race to optimize system efficiency and organizational effort. Their on-road problems were confined to only three flat tires.

**TABLE 13.4**
**SOLAR CARS, AVERAGE SPEED AND**
**MECHANICAL POWER, PREDICTED SPEED**

| | Avg. Speed km/h | Required W@/ Speed* | Speed W/449W km/h | Speed Up 1% Grade, 449W km/h | Speed Up 2% Grade, 449W km/h |
|---|---|---|---|---|---|
| General Motors USA | 66.92 | 921. | 47.11 | 31.34 | 16.49 |
| Ford Australia | 44.49 | 449. | 44.49 | 30.05 | 21.42 |
| Ingenieurschule Biel | 42.95 | 424. | 44.19 | 29.50 | 20.55 |
| Australian Geographic | 36.90 | 451. | 36.82 | 24.80 | 17.40 |
| Darwin Institute of Technology | 31.48 | 382 | 34.72 | 23.00 | 16.28 |
| Chisholm Institute | 30.60 | 358. | 34.70 | 22.33 | 15.34 |

*Net mechanical power required to propel the solarmobile on the level with no wind.

**Ford Australia**

(As Table 13.4 shows) Ford's tilting array vehicle would have been competitive to GM's if the solar arrays were equal. The Ford vehicle had better stability in cross winds, but a higher wind resistance by using four exposed wheels. To allow their array to tilt further, the Ford design used a rear wheel track that was narrower than the front—meaning the rear wheels could not draft behind the front.

The Ford team's performance was affected by motor and transmission problems and an unfortunate battery management system error that failed to properly charge their silver-zinc batteries. The Ford team were not pleased to discover that they had left Darwin with a battery set that was only 50% charged. This omission limited their performance and prevented them from staying closer to GM and better weather.

The inclement weather that plagued all but the GM team also created another problem for the Ford team. Water leaking into their array caused a serious short-circuit that had to be repaired. The Ford car was also caught in a severe hail storm that damaged the car's array and reduced its electrical output by about 10%.

**Ingenieurschule Biel**

The Swiss lived up to their reputation for manufacturing machines that are both reliable and precise. Unfortunately, a wrong turn in Alice Springs and an unseen Australian "give way" sign sent the *Spirit of Biel* into the side of an oncoming car and prevented the Swiss from claiming second place. After the accident, which severely damaged the car's steering and array support structure, the Swiss team demonstrated their organizational abilities to effect repairs in less than six hours. The accident happened early in the morning, which gave the Ford team a four-hour advantage. (They were two hours behind at the time of the accident.) The race would certainly have been closer if the accident had occurred closer to the day's official finish time. Apart from the incident, the Swiss team had no breakdowns and a design that minimized air resistance by placing the rear wheel in the center. According to Dr. Kyle's prediction model, with the same weight as Ford (274 kg [604 lb]) and equal power, the Swiss car would have been slightly faster (44.82 km/h vs. 44.19 km/h [27.85 mph vs. 27.46 mph]).

### Australian Geographic

The *Team Marsupial* car also exhibited a high degree of reliability with the exception of a few minor wheel problems. In terms of cost and performance, this car was probably the best in the race. The car utilized wheel fairings, low-rolling-resistance tires, and an excellent exterior finish. However, the aerodynamic efficiency suffered from aerodynamic separation behind the cab and rear wheel fairings.

### Darwin Institute of Technology

The *Desert Rose* is an example of how minor details can make a significant difference in the results. Although the car was almost identical to the Biel car, the results were poorer due, most probably, to time pressures. The team was still working to finish the car when the race began. Their performance could have been improved with the use of lower-rolling-resistance tires. By using bicycle-type tires as opposed to motorcycle tires and the same power as the AG car (451 watts), the Darwin entry's predicted speed, according to Dr. Kyle's model, would have been 38.97 km/h vs. 36.90 km/h (24.22 mph vs. 22.93 mph) for the AG car. Even with their low available power (382 watts), the Darwin car would have been faster (37.5 km/h vs. 36.8 km/h [23.3 mph vs. 22.87 mph]) if its weight were equal to the Biel car and replacement bicycle tires were fitted.

### Chisholm Institute of Technology

Perhaps the most unusual design in the race was Chisholm's *Desert Cat* (Figure 13.15). The double-hull catamaran design was relatively heavy at 385 kg (849 lb), had a high aerodynamic drag, poor electronics, and wheel and tire problems. However, a very large solar array negated most of its shortcomings. The array's output was second only to GM's, and, by improving in some of the other areas, the *Desert Cat* would probably have equaled or surpassed the Darwin Institute vehicle's speed.

Chisholm's array design with large side panels meant that maximum power was reached between 10 A.M. and 2 P.M. In theory, their array would generate about 6% more energy than a tilting array in clear sunlight and even more in diffuse lighting conditions. The future of this type of design is doubtful, however, unless good aerodynamics and electronics simplicity can be attained. Intricate array shapes, as in the Chisholm car, serviced by a series of Maximum Power Point Tracking (MPPT) devices are complicated and expensive.

*Figure 13.15.* Desert Cat, *Chisholm Institute of Technology, Australia. (Bruce McCristal)*

### Semiconductor Energy Laboratory (SEL)

The *Southern Cross* team deserves mention as the final team to cross the finish line. This assertion is not to denigrate their effort, but to congratulate them for their commitment and perseverance in finishing the race with a vehicle that was not designed to be fast but to be the only one to use amorphous cell technology. The company's entry was a statement on the potential use of the cheaper but less-efficient amorphous (crystal) technology...

### Conclusions

A number of conclusions based on the data presented here are offered...Some of these conclusions are subjective and based on the observations presented by a number of sources, including the teams themselves.

a)  Reliability was the most important factor in winning. Part of this strategy is completing construction well before race time and thorough testing of the vehicle to identify and solve potential problems. The GM, Swiss, and Australian Geographic (AG) solarmobiles were all thoroughly tested and very reliable.

b)  The order of importance of the critical physical design factors was: 1) solar array power, 2) aerodynamics, 3) weight, and 4) rolling resistance.

c)  Race strategy was also an important factor. This includes careful analysis and planning of driver training, organization, logistics, and repair capabilities. The GM team epitomized this process.

d)  With equal equipment, the top three cars would have been very close. The GM runaway victory can be mostly attributed to the use of the ultra-efficient gallium-arsenide cells. Future racers should consider a fresh approach to design. Quoting Dr. Kyle, "the definitive design has not appeared."

e)   Race organizers may wish to consider alterations to existing rules to increase competitiveness and limit cost. These (rules) would include restricting the types of solar cells and batteries and also to control the capacity of battery storage.

f)   Due to the large variation in performance among vehicles, the field was strung along the highway over 1000 km (621 miles), making communications a problem. Future races may initiate a staggered start or a race held in stages to keep the field closer together. This approach would have organizational as well as media benefits.

A final observation: the event evoked a deep sense of cooperation among teams. Competition was based on each team helping themselves and their competitors to achieve their maximum potential. Several teams loaned spare equipment to keep opposing vehicles in the race. If there is one concept that should not change in future races, it is this very worthwhile attitude among racers.

The first transcontinental solarmobile race was a melting pot of technology, ideas, and innovations that will have significant spin-offs in the solar, automotive, and battery industries long after the newspaper headlines fade. Whether the accomplishments of this World Solar Challenge (Race) will be surpassed is an admirable goal for the 1990 racers. This competition and future races will be viewed as a barometer of progress in renewable energy and transportation technologies.

Dr. Kyle's final summary appears in Table 13.5.

Figure 13.16 shows the GM *Sunraycer*, at the finish line of the 1987 World Solar Challenge Race. The perpetual trophy is shown in Figure 13.17.

In the future for the Trans-Australian World Solar Challenge Race, similar to the *Tour de Sol,* the promoters may wish to establish limits on both the emissive power from the solar arrays and the energy capacity of the onboard battery. This will encourage reasonable types of vehicles, applicable with small modification to civilian use, rather than entries which, through successive races, evolve into specialized monstrosities. Almost any observer views these contests as beneficial in encouraging the exploitation of this age-old energy source for silent and pollution-free transportation. The principle appears to have a definite future. Needless to say, the concept also is applicable to the twenty-first century, solar-power, propeller-driven, superconducting-magnet-equipped, levitating vehicle described elsewhere. In closing, an elegantly prepared book privately printed by General Motors Corporation lavishly describes and pictures this race, with color illustrations.[7]

The next chapter discusses the American *Tour de Sol* of 1992. High-level finishers of this race were encouraged to participate in the World Solar Challenge Race in Australia the following November.

## TABLE 13.5
## SIGNIFICANT PARAMETERS BY DR. KYLE
## 1987 AUSTRALIANWORLD SOLAR CHALLENGE RACE

| Place Car | Wt. Less Driver (lb) | Average Race Speed (mph) | Qualifying Speed on Batteries (mph) | Best Daily Distance (miles) | Drag Area CdA (sq.ft)* | Average Panel Solar Exposure (6 A.M.– 7 P.M.)** | Predicted Speed @ Identical Si Cells and Electrical System*** |
|---|---|---|---|---|---|---|---|
| 1. Sunraycer GM USA | 397. | 41.6 | 70.2 | 376. | 1.53 | 90.%[a] | 36.9 |
| 2. Sunchaser Ford Australia | 417. | 27.6 | 49.7 | 279. | 1.94 | 100.[b] | 36.59 |
| 3. Biel Swiss | 448. | 26.7 | 44.1 | 303. | 1.87 | 100.[b] | 36.58 |
| 4. Marsupial Australian Geographic | 562. | 22.9 | 52.8 | 263. | 2.91 | 92.[a] | 30.49 |
| 5. Desert Rose Darwin Institute | 551. | 19.6 | xx | 237. | 2.72 | 100.[b] | 29.97 |
| 6. Desert Cat Chisholm Institute | 662. | 19.0 | 41.6 | 231. | 3.47 | 106.[c] | 29.11 |
| 7. Solar Res. Australia | 389. | 15.9 | 26.1 | 170. | 2.6 | 88.[d] | 31.89 |
| 8. Star USA Crowder College | 270. | 15.4 | 32.3 | 172. | 3.71 | 93.[b] | 30.95 |
| 9. Solectria IV MIT USA | 301. | 15.2 | xx | 180. | 2.6 | 100.[b] | 34.62 |
| 10. Mana La Hawaii USA | 573. | 13.4 | 57.8 | 236. | 2.7 | 95.[a] | 28.59 |
| 11. Chariot Denmark | 484. | 12.4 | 19.3 | 147. | 3.0 | 96.[e] | 26.07 |
| 12. Phoebus II Japan | 609. | 12.2 | 30.4 | 134. | 3.9 | 93.[c] | 26.06 |
| 13. Photon F. Morphett | 627. | 9.9 | 18.0 | 104. | 4.4 | 88.[d] | 26.03 |
| 14. Southern Cross SEL Japan | 627. | 6.7 | 12.4 | 72. | 4.3 | 96.[b] | 26.29 |

[a] Fixed curved panel.
[b] Tilting panel.
[c] Fixed flat panel with fixed sides.
[d] Fixed flat panel.
[e] Fixed flat panel with tilting sides.
* The drag area is the aerodynamic drag coefficient Cd times the effective frontal area A. It can be considered as the equivalent drag of a flat plate facing the wind. The smaller the area, the lower the drag. This was the only measure of aerodynamic drag available for all the cars.
** Each type of panel is capable of intercepting an amount of solar energy, depending on the average projected area facing the sun during the day. For reference, a tilting panel is considered 100% effective, and the others are compared to this. All of the panels could be tilted toward the sun when the cars were stationary, and therefore any advantage occurs only during racing hours from 8 A.M. to 5 P.M. The reference figure is the average for the entire time the panels may be exposed to the sun from 6 A.M. to 7 P.M.
*** Using a mathematical model which includes the characteristics of each car including power to the drive wheels, weight, aerodynamic drag, and tire rolling resistance, the predicted speed on the level with no wind is shown. For this calculation, all the cars have silicon solar panels and electrical systems of identical efficiency. The calculation shows that the first six places would have remained unchanged, regardless of the sophistication of the equipment used. Essentially, the design factors of aerodynamic drag, weight, tire rolling resistance, and panel design determined the placing rather than the type of solar cells or the efficiency of the electrical system.

*Figure 13.16. Finish World Solar Challenge race of 1987, General Motor's* Sunraycer, *winner. (Bruce McCristal, General Motors)*

# Other Vehicles

Photos of other vehicles that participated in the race but were not shown in the preceding pages are included here as Figures 13.18 through 13.21.

In addition, the following is a list of world class solarmobiles:

1. *Photon Flyer*—Australia
2. Rolf Disch—Germany
3. Syed Attique Shafaat—Pakistan
4. *Chariot of the Sun*—Denmark
5. Nippon TV, Leyton—Japan
6. Hoxan Corporation—Japan
7. Chisholm Institute of Technology—Australia
8. John Paul Mitchell Systems—U.S.A.
9. Australian Geographic—Australia
10. Ford Motor Company—Australia
11. Solar Resource Syndicate—Australia
12. General Motors Corporation—U.S.A.

*Figure 13.17.  Perpetual Trophy, Broken Hill Associated Smelters, Donor.
(Christine M.  Burnup, Pasminco Ltd., Melbourne, Australia)*

*Figure 13.18.*  Phoebus II, *Hoxan, Japan.  (Bruce McCristal, General Motors)*

*Figure 13.19.* Solectria IV, *MIT. (Bill Tuckey)*

*Figure 13.20.* Chariot of the Sun, *Denmark. (Bruce McCristal, General Motors)*

*Figure 13.21. A RADAR means of determining speed to establish pole position. (Bill Tuckey)*

# References

1.  *Smithsonian*, **18**, 11, 1988, pp. 48–58. See also: "Pentax World Solar Challenge Official Programme," furnished courtesy of Tokyo Eizosha Co. Ltd., 8-13, Azabu Juban 1-Chome, Minato-ku, Tokyo, Japan.
2.  Wakefield, Ernest H., *History of the Electric Automobile*: *Battery-Only Powered Cars,* Society of Automotive Engineers, Warrendale, PA, 1994, pp. 331-356.
3.  Asahi Optical Ltd., 36-9, Maeno-Cho, 2-Chome, Itabashi-ku, Tokyo 1744, Japan.
4.  *The Advertiser,* Adelaide, South Australia.
5.  *The Darwin Northern Territorian*, Darwin, Northern Territory, Australia.
6.  Ingersoll, E.P., *The Horseless Age*, Vol. 1, No. 1, November 1895, pp. 1–2.
7.  Tuckey, Bill, *Sunraycer's Solar Saga*, The Berghouse Floyd Tuckey Publishing Group, P.O. Box 303, Gordon, NSW, 2027 Australia, 1987.
8.  Fries, Peter. O., "The 1987 World Solar Challenge: A Technical Report for the New Energy Development Organization of Japan," Renewable Energy Services, P.O. Box 33, St. Lucia, Queensland, 4067 Australia; report prepared for Energy Promotions, 1697 Pittwater Rd., Mona Vale, NSW (a division of Flipper Scow Pty. Ltd.), with funding provided by New Energy Development Organization, January 1988.
9.  Kyle, Chester R., "How Design Factors Affect Solar Car Race Performance," SAE Paper No. 880728, Society of Automotive Engineers, Warrendale, PA, 1988.

# CHAPTER 14

# The American
# *Tour de Sol*—1992

## Background

Sunrayce USA (details omitted here because of space limitation) is held every three years and focuses specifically on college and university-built solar-dominant electric race cars. In contrast, the annual American *Tour de Sol* is more broadly gauged. It emphasizes solar-dominant race and solar-assist commuter cars and, indeed, solar-assist work vehicles such as vans and pickup trucks built by individuals, institutions, and industry. Both competitions, as well as the Swiss *Tour de Sol* (details omitted here for lack of space) and the Trans-Australian World Solar Challenge Race, serve to enhance science, to heighten interest in utilizing energy from the sun, and to introduce the use of solar energy into personal transportation.

In 1987, the GM *Sunraycer*, discussed in Chapter 13, piqued interest in personal transportation by reaching speeds of 70 mph (113 km/h) and consumed 30 watt-hours per mile while traveling at 55 mph (88 km/h) on energy from only the sun, a figure equal to approximately 1000 mpg (425 km/l) of gasoline. Readers of *History of the Electric Automobile: Battery-Only Powered Cars*[1] will notice how solar-electric car racing described in the present book, *History of the Electric Automobile: Hybrid Electric Vehicles,* nearly apes the electric car experiences of the last decade of the nineteenth and first part of the twentieth centuries as reported in Ref. 1. People do not seem to change—only their technologies change! You can imagine that the twenty-first century can be the era when energy from the sun may strongly influence the mechanical, chemical, and electrical services of people. Economies already use energy from the sun in water and wind power and in drying and agriculture. For agriculture, Professors David and Marcia Pimental of Cornell University have written that the farmer—

by plowing, planting, and harvesting—supplies only 10% of the energy required to deliver crops. The sun supplies the balance.[2] After all, the solar energy from a zenith sun falling each hour on a singles tennis court that measures 78 × 27 ft (24 × 8 m) is equivalent to the energy contained in approximately 5 gallons (19 liters) of gasoline. What a pity for mankind not to use solar energy even more effectively.[a,3]

Nancy Hazard, associate director of the Northeast Sustainable Energy Association, relates another innovation: how solar racing has moved the female gender into mechanical and electrical engineering of solar-electric-based transportation. She writes[4]:

> It wasn't one of the original intentions of the American *Tour de Sol* solar and electric vehicle race, but race officials are pleased with the development. A larger number of women are involved in this year's event than ever in its four-year history. What was surprising was the large number of women involved in the design, engineering, and construction of electric and solar-electric vehicles planning to compete in this annual event. Nearly every major team registered for this year's competition included women involved in key areas of design and construction. In the past, automotive technology has not attracted a significant number of women in this country.
>
> The American *Tour de Sol* is a major electric vehicle competition geared toward the development of practical, pollution-free, commuter-oriented vehicles. Entrants in the American *Tour de Sol* typically come from one of three backgrounds: individual engineers, college teams, or corporate entries. Some of the most successful corporate entries in past years have been produced by Massachusetts-based Solectria Corporation (Arlington, Massachusetts). Solectria's president and cofounder Anita Rajan explains the appeal of this energy technology. "There are so few women involved at present in internal combustion engine technology, they just don't have much impact. But in the development of electric and solar car technology, women's message is being heard: women are bright, energetic, and can have an impact on emerging electric vehicle technology." Ms. Rajan has been involved in this field for four years since (being graduated from) MIT with a degree in electrical engineering.

Another outstanding woman in the development of solar-electric vehicles has been Molly Brennan, Manager of Customer Programs, General Motors Technical Center. Ms. Brennan holds a bachelor's degree in computer science and humanities from Michigan State University. She also attended Oxford University in England for two years as a Rhodes Scholar and received a degree in philosophy, politics, and economics. Most recently, Ms. Brennan was Observer-Manager of the GM Sunrayce USA—1990. In 1987, she was one of the drivers of the GM *Sunraycer* when it competed in the first World Solar Challenge Race in Australia. Ms. Brennan is shown in Figure 14.1.

---

[a] Ayers Rock Resort, almost in the center of Australia (P.O. Box 46, Yulara Northern Territory, 0872 Australia), generates a peak of 250 kW from solar cells.

*Figure 14.1. Molly Brennan at the wheel of the GM Sunraycer. (Bruce McCristal)*

Another notable woman is Nancy Hazard, a descendant of Commodore Oliver Hazard Perry of the 1812 Battle of Lake Erie fame. She should be included in this pantheon of female participation in solar-related car racing, being associate director of the Northeast Sustainable Energy Association and one of the most articulate members in promoting the American *Tour de Sol.*

Ms. Hazard continues:[4]

> Another first this year (in the American *Tour de Sol*) is an entire women's race team from the Ethel Walker School in Connecticut. About a dozen young women from an honors physics class form the nucleus of the team, but dozens of other students have become involved in support roles. Their entry is one of the typically more high-tech-looking designs which are found in the Racing category of the American *Tour de Sol.* Racing cars are usually futuristic-looking one-person vehicles built with twentieth-century composite materials. By comparison, vehicles entered in the Commuter categories often use the body of a production car and appear more ordinary except for the quiet (motor) and the lack of exhaust.

This year's American *Tour de Sol* runs from Albany, New York, through western Massachusetts, the length of Connecticut including Hartford, and ends in the Boston area at the Museum of Transportation in Brookline. The race leaves Albany on Monday, 18 May, and finishes at the museum about midday on Friday, 22 May. Two day-long exhibits of the race vehicles are planned for Sunday, 17 May, at the Empire State Plaza, and for Saturday, 23 May, on Boston's City Hall Plaza...Race organizers expect approximately 40 vehicles to be registered in this year's event. Major sponsors for the 1992 American *Tour de Sol* include the U.S. Department of Energy through Argonne National Laboratory, the New York State Energy Research and Development Authority, the New York State Energy Office, and the New England Electric System.

# The American *Tour de Sol* Race

In her release about the race,[4] Ms. Hazard writes:

A solid week of sunshine set the tone for the fourth annual American *Tour de Sol* solar and electric vehicle race. Organizers say this year's event far exceeded their expectations in all respects...Thirty-five vehicles traveled over 250 miles (400 km) from Albany, New York, to cross the finish line 22 May at Boston's historic Museum of Transportation.

Organized by the Northeast Sustainable Energy Association (NESEA) of Greenfield, Massachusetts, the American *Tour de Sol* is the premier championship race devoted to the development of commuter-type vehicles. At the same time, the race offers a showcase for some of the country's most successful high-tech racing vehicles being developed by corporations, schools, and individuals. Thousands of people came to see the race, and they were very impressed with quality of the cars, which included several models commercially available. It is exciting to see the rapid development of electric vehicles over the past four years, which has been stimulated by legislation that will bring them to market in the second half of the decade as a (partial) solution to urban air pollution, foreign oil dependence, and global warming.

After rigorous safety testing in Albany, which included data collection, visual inspections, and stability, steering, and braking tests, the sixteen commuter cars, twelve racing cars, and five open-category vehicles settled into the daily routine of the early morning drivers' meeting, followed by a morning and afternoon drive, each of which ended at an educational display of the cars. At evening displays, cars were encouraged to drive extra miles to demonstrate range and to gain time credits. Stiff time penalties were levied on cars that failed to complete the required miles. Each day had its unique challenges, such as a hill climbing contest at Avon Mountain, and range and efficiency tests held at the Thompson Speedway in Thompson, Connecticut.

This year, eight cars competed in a newly created category: the American Commuter category. It was created to accommodate heavier cars, such as conversions. Unlike the *Tour de Sol* Commuter category, which restricts the battery capacity and the output of the photovoltaic array, there are no restrictions on American Commuter cars, except that (each)...must carry at least one passenger.

Solectria Corporation's *Flash* and *Force* placed first and second, respectively, in the lighter weight *Tour de Sol* Commuter category, followed by *Solar Tech II*, built by the New England Institute of Technology in Warwick, Rhode Island. Solectria Corporation, founded by MIT graduates three years ago, offers two electric cars for sale. The cars, built to order, have been sold to a number of early adopters as well as electric utility companies. An efficient motor and drive-train system, developed by the company, and nickel-cadmium batteries also enabled Solectria to capture the prize for Most Efficient and for Greatest Range, after driving 100.2 miles (161 km) on a single battery charge. (Solectria's *Force GT* is illustrated in Figure 14.2.)

"While the Commuter category cars demonstrate practicality, the solar racing cars beautifully demonstrate the potential of a sustainable transportation system powered by electricity generated from the sun," states Dr. Robert Wills, codirector of the American *Tour de Sol.* In the racing categories, the Conval High School entry, *Sol Survivor II* (Figure 14.3), of Peterborough, New Hampshire, took first place in the *Tour de Sol* Racing category, over college and university-built cars. They were closely followed by entries built by the University of Lowell and New Hampshire Technical Institute (NHTI) of Concord. Cross

*Figure 14.2. Solectria's* Force GT. *(Nancy Hazard)*

Continental Racing category honors went to Villanova University, Virginia Polytechnic Institute's *Solaray II* (Figure 14.4), and Drexel University (each of which participated in the GM Sunrayce USA—1990). Wills commented further: "The big change in the American *Tour de Sol* this year was that more cars were traveling at normal road speeds, completing the 50-mile (80-km) daily race route, and driving extra laps each day at these speeds without a problem."

A last minute entry from the New Hampshire Technology Institute (NHTI) was the completely rebuilt *Sungo*. Originally built around a converted *Yugo*, the *Sungo* has a massive aluminum frame and composite body that combine strength and light weight. This combination, together with dual Solectric (permanent magnet synchronous motors) won NHTI the Avon, Connecticut, hill climb competition, as well as the most efficient student-built entry among the *Tour de Sol* Commuters.

Another interesting vehicle was the converted *S-10* Chevrolet pickup from Solar Car Corporation of Melbourne, Florida, shown in Figure 14.5. The 3500-lb (1590-kg) *Electro-Chevy,* sponsored by electric utility companies, performed excellently against the other smaller and lighter passenger cars in the American Commuter category. The Chevy *S-10*

*Figure 14.3. The Conval High School team's* Sol Survivor II.
*(Nancy Hazard. Photo by C. Michael Lewis)*

conversion is a natural choice for fleet owners and anyone interested in trying out electric vehicle technology, as the vehicle is already designed to carry the extra weight that a battery-pack imposes, and Solar Car's design eliminated the need for any batteries in the passenger compartment. Solar Car Corporation offers both fully finished converted vehicles and kits, so that users may convert their own vehicles.

The U.S. Department of Energy, through Argonne National Laboratory, awarded $7500 to top-placing student-built commuter cars to "stimulate young people to become involved in technical careers," according to Phil Patterson, the U.S. Department of Energy Office of Transportation Technology students' competitions director. The St. Johnsbury Academy *Electric Hilltopper* received first place in the American Commuter category and also received the prize for Greatest Range (Figure 14.6). The St. Johnsbury Academy *Jewel* placed second, followed by the Waterbury (Connecticut) State Technical College entry. In the *Tour de Sol* Commuter category, the *Solar Tech I* and *II* entries (Figure 14.7) from New England Institute of Technology in Warwick, Rhode Island, took first and second place, followed by the Unatego High School entry from Otego, New York.

*Figure 14.4. The Virginia Polytechnic Institute* Solaray II. *(Nancy Hazard. Photo by Sandy Hill)*

*Figure 14.5. Solar Car Corporation solar-assist S-10 Chevrolet.
(Douglas Cobb, Solar Car Corporation)*

*Figure 14.6. The St. Johnsbury Academy* Electric Hilltopper *and* Electric Jewel. *(Nancy Hazard)*

Open-category contestants were a delight, with several one-person vehicles, a motor-cycle, and a solar-powered recumbent tricycle by Bruce Meland, illustrated in Figure 14.8. Again, a high-school entry took the day: Cato-Meridian's *C-M Sunpacer* (Figure 14.9) took first place, followed by MIT's *Aztec* (Figure 14.10) and Central Connecticut State University's electrified motor cycle, the *Envirocycle* (Figure 14.11).

For the first time ever, the American *Tour de Sol* stopped at a race track—the Thompson Speedway in Thompson, Connecticut. Cars competed for a range prize, and efficiency data was collected on the cars at various speeds, which was added to data collected during the rigorous pre-race testing. The efficiency of electric cars is measured in watt-hours per mile rather than miles per gallon. Surprising efficiencies were demonstrated by the cars, with top runners driving a mile on less energy than running a 100-watt light bulb for an hour. If translated to pure energy use, this is comparable to a gasoline car running 500 miles on a single gallon of gasoline (213 km/l). Data collection, which was made possible by a grant from the U.S. Department of Energy, will be published in NESEA's quarterly magazine, the *Northeast Sun*.

*Figure 14.7. The New England Institute of Technology Solar Tech II.*
*(Nancy Hazard. Photo by C. Michael Lewis)*

*Figure 14.8. Solar-powered recumbent tricycle. (Bruce Meland, Solarland, Bend, Oregon)*

*Figure 14.9. The Cato-Meridian team's* Sunpacer. *(Nancy Hazard)*

*Figure 14.10. The MIT Aztec. (Nancy Hazard. Photo by Craig Line)*

*Figure 14.11. The Central Connecticut State University Envirocycle.*
*(Nancy Hazard. Photo by Frank Poszlusny)*

"Race logistics ran incredibly smoothly, as over 100 volunteers assisted with everything from timing to traffic control," said Kristen Walser, volunteer coordinator. "Communications support from Southern New England Telephone (Company) and Radio Resources (Company) were invaluable, as was support from local schools, towns, and groups such as Peoples' Action for Clean Energy, the Mansfield Common Ground, and Green Pastures Power Company." A number of exhibitors joined the cars at displays along the route, including Argonne National Laboratory, New England Electric System's *Energy Van,* Ben and Jerry's Traveling Show, General Motors, and Sunnyside Solar, to name a few. Miniature solar car races were also held at a number of stop-over locations with the assistance of the U. S. Department of Energy and the Society of Automotive Engineers.

"National publicity was excellent, with CNN, *The New York Times*, *The Boston Globe*, AP, and UPI carrying the event, and over 50 media outlets around the world asked for daily updates," said Jack Groh, of Media One, who handled the public relations. "It is gratifying to see how the news media and the general public have begun to pay closer attention to this event. I believe this indicates that the electric vehicle industry will soon be coming of age in this country. "

Tables 14.1 and 14.2 show the results of the American *Tour de Sol*—1992.

# Vehicle Range Increases with Time

One measure of the value of competitions such as the American *Tour de Sol* is recording of the enhancement with time in both range and efficiency of electric vehicles. As Nancy Hazard, associate director of Northeast Sustainable Energy Association, has written[5]:

Range, or the distance a car can drive before refueling, is considered by many to be the Achilles' heel of electric vehicles. But range records have increased from 35 miles to over 200 miles (56 to 321 km) in six short years, as measured at the American *Tour de Sol.*

We have measured dramatic increases in both range and efficiency of electric vehicles (EVs) over the past six years at NESEA's American *Tour de Sol* and (think range is an ever smaller) barrier to the commercialization of EVs...Increased range is a function of increasing the "gas tank"—the onboard storage of energy, as well as increasing the vehicle efficiency—the energy it takes to move the car down the road...As shown in the graph (Figure 14.12), the range of EVs using lead acid batteries as their energy storage system has increased from 35 to 142 miles (56 to 228 km). The 142-mile (228-km) record was set by a Fiat converted to electric propulsion by Bolton High School in Bolton, Connecticut. When advanced energy storage systems are used, ranges of over 150 miles (241 km) are common. In 1994, the *Solectric Force*, a Geo *Metro* converted by Solectria Corporation of Wilmington, Massachusetts, drove 214 miles (344 km) using a nickel-metal-hydride battery built by Ovonic Battery of Troy, Michigan. Ovonic plans to put its battery on the market by 1998, and other energy storage systems with similar capabilities are in the wings.

**TABLE 14.1**

**1992 AMERICAN *TOUR DE SOL* RESULTS—RACING CARS**

| Overall Position | Adjusted Time (Hours:Minutes) | Total Miles | Range | Efficiency, Watt-Hours/ Mile @ 25 mph | Vehicle Name | Team |
|---|---|---|---|---|---|---|
| | | | | | *Tour de Sol Racing* | |
| 1 | 6:44 | 300.0 | 80.3 | 36.9* | *Sol Survivor II* | Conval High School Peterborough, NH  Paul Waterman |
| 2 | 8:26 | 240.9 | 32.3 | — | *Sunblazer* | University of Lowell Lowell, MA  John Duffy |
| 3 | 10:42 | 306.6 | 66.2 | 60.6 | *Suntech* | New Hampshire Technical Institute (NHTI) Concord, NH  Thomas P. Hopper |
| 4 | 13:08 | 297.2 | 60.0 | — | *Sunvox IV* | Dartmouth College Hanover, NH  Doug Fraser |
| 5 | 43:07 | 171.9 | 38.6 | 64.2 | *Sunvox I* | Dartmouth College Hanover, NH  Doug Fraser |
| 6 | 45:46 | 122.0 | 32.3 | — | *Husky Solar Electric* | Northeastern University Boston, MA  Greg Morin |
| 7 | 60:19 | 91.9 | — | — | *Sol Dancer* | Ethel Walker School Inc. Simsbury, CT  Pam Akiri |
| 8 | 68:14 | 92.7 | — | 167 | *Sun Lion* | Trenton State College Trenton, NJ  Norman Asper |
| | | | | | **Cross Continental Racing** | |
| 1 | 7:04 | 293.4 | 63.6 | 65.3 | *Wild Solarcat II* | Villanova University Villanova, PA  Gary Nevard |
| 2 | 8:80 | 315.1* | 67.8 | — | *Solaray II* | Virginia Polytechnic Institute Blacksburg, VA  Charles Hurst |
| 3 | 18.41 | 248.6 | 81.9 | — | *Sun Dragon II* | Drexel University Philadelphia, PA  John Kodsi |
| 4 | 73.23 | 35.7 | — | — | *Sun Quest* | Queens Solar Vehicle Kingston, Ontario  Andrew Marchant |

\* Range and efficiency prizes.

TABLE 14.2
1992 AMERICAN *TOUR DE SOL* RESULTS—COMMUTER AND OPEN

| Overall Position | DOE Prize* | Adjusted Time (Hours:Minutes) | Total Miles | Range | Efficiency, Watt-Hours/ Mile @ 25 mph | Vehicle Name | Team |
|---|---|---|---|---|---|---|---|
| | | | | | American Commuter | | |
| 1 | — | 3:35 | 407.6 | 100.2** | 147.8** | *The Force GT* | Solectria Corporation Arlington, MA James Worden |
| 2 | 1 | 4:12 | 393.6 | 96.0*** | 209.4 | *The Electric Hilltopper* | St. Johnsbury Academy St. Johnsbury, VT Bruce Burk |
| 3 | 2 | 4:59 | 354.8 | 77.2 | 204.5 | *The Electric Jewel* | St. Johnsbury Academy St. Johnsbury, VT Bruce Burk |
| 4 | — | 5:20 | 331.2 | 90.8 | 206.7 | *Electro Chevy* | Solar Car Corporation Melbourne, FL Bob Adams |
| 5 | — | 5:59 | 337.0 | 66.8 | 338.2 | *Poetry in Motion* | Boston, MA Al Hutton |
| 6 | — | 6:57 | 297.5 | 63.6 | 187.0 | *E96* | Vermont Electric Car Montpelier, VT Paul Scheckle |
| 7 | 3 | 7:09 | 333.9 | 87.6 | 354.6 | *Kineticar* | Waterbury State Technical College Waterbury, CT Prof. D.C. Narducci |
| 8 | — | 11:41 | 259.0 | 48.5 | 305.4 | *VW Conversion* | Staten Island, NY Scott Isgar |

**TABLE 14.2 (continued)**

### Tour de Sol Commuter

| Overall Position | DOE Prize* | Adjusted Time (Hours:Minutes) | Total Miles | Range | Efficiency, Watt-Hours/ Mile @ 25 mph | Vehicle Name | Team |
|---|---|---|---|---|---|---|---|
| 1 | — | 5:16 | 345.1 | 84.5** | 74.8** | *The Flash* | Solectria Corporation Arlington, MA James Worden |
| 2 | — | 5:59 | 289.5 | 73.5 | 139.9 | *The Force* | Solectria Corporation Arlington, MA James Worden |
| 3 | 1 | 23:05 | 239.2 | 41.7 | 246.5 | *Solar Tech II* | New England Institute of Technology (NEIT) Warwick, RI James Adams |
| 4 | 2 | 23:47 | 217.1 | 39.6 | — | *Solar Tech* | New England Institute of Technology (NEIT) Warwick, RI James Adams |
| 5 | 3 | 31:28 | 200.5 | 49.5 | — | *Solar Bullet* | Unatego High School Otego, NY Paul Agoglia |
| 6 | — | 32:31 | 179.6 | 37.0 | 192.8 | *'Lectric Lizzie* | Champlain College Burlington, VT Mark Pouliot |
| 7 | — | 38:20 | 154.3 | 33.2 | — | *S-CAR-GO* | Delta College University Center, MI Jack Crowell |
| 8 | — | 40:07 | 160.6 | 57.9 | 88.36*** | *Sungo* | New Hampshire Technical Institute (NHTI) Concord, NH Thomas P. Hopper |

## TABLE 14.2 (continued)

| Overall Position | DOE Prize* | Adjusted Time (Hours:Minutes) | Total Miles | Range | Efficiency, Watt-Hours/ Mile @ 25 mph | Vehicle Name | Team |
|---|---|---|---|---|---|---|---|
| | | | | | Open | | |
| 1 | 1 | 10:26 | 282.2 | 66.2 | — | C-M Sunpacer | Cato-Meridian High School Cato, NY Earl Billings |
| 2 | 2 | 12:06 | 325.3 | 80.3 | 59.8 | Aztec | MIT Solar EV Club Cambridge, MA David Hampton |
| 3 | 3 | 21:03 | 261.6 | 56.3 | 130.0 | Envirocycle | Central Connecticut State University New Britain, CT Dr. John Wright |
| 4 | — | 32:20 | 216.2 | 60.5 | 145 | S.E.T.S. Sunrunner | S.E.T.S. Racing Enosburg Falls, VT Ed Gaudette |
| 5 | — | 58:23 | 106.7 | 33.9 | — | Moves with the Sun | Jim Coate Belmont, MA Jim Coate |

\* $7,500 U.S. Department of Energy (DOE) prize money, through Argonne National Laboratory, for the best student-built commuter cars.

\*\* Range and efficiency prizes.

\*\*\* U.S. Department of Energy (DOE) prizes for best range and efficiency prizes for student-built commuter cars.

Miscellaneous DOE prizes:     Most Improved = Delta College.     Teamwork and Team Spirit = New England Institute of Technology.

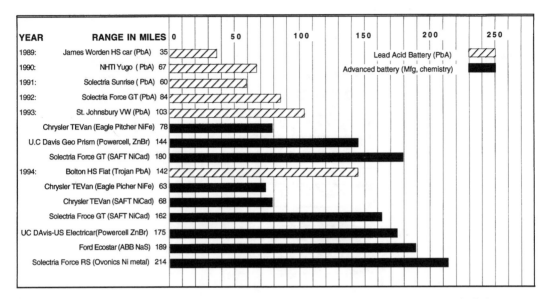

| YEAR | RANGE IN MILES | |
|---|---|---|
| 1989: | James Worden HS car (PbA) | 35 |
| 1990: | NHTI Yugo ( PbA) | 67 |
| 1991: | Solectria Sunrise ( PbA) | 60 |
| 1992: | Solectria Force GT (PbA) | 84 |
| 1993: | St. Johnsbury VW (PbA) | 103 |
| | Chrysler TEVan (Eagle Pitcher NiFe) | 78 |
| | U.C Davis Geo Prism (Powercell, ZnBr) | 144 |
| | Solectria Force GT (SAFT NiCad) | 180 |
| 1994: | Bolton HS Fiat (Trojan PbA) | 142 |
| | Chrysler TEVan (Eagle Picher NiFe) | 63 |
| | Chrysler TEVan (SAFT NiCad) | 68 |
| | Solectria Froce GT (SAFT NiCad) | 162 |
| | UC DAvis-US Electricar(Powercell ZnBr) | 175 |
| | Ford Ecostar (ABB NaS) | 189 |
| | Solectria Force RS (Ovonics Ni metal) | 214 |

Lead Acid Battery (PbA)
Advanced battery (Mfg, chemistry)

*Figure 14.12. Electric vehicle range increase measured at NESEA American* Tour de Sol *over six years (1989 to 1994). (Nancy Hazard, Northeast Sustainable Energy Association)*

Besides increased energy storage, Solectria's success was made possible by technical advances that increased the efficiency of its vehicle as demonstrated by the efficiency graph (Figure 14.13). Here we see that in two years, Solectria increased the efficiency of a converted Geo *Metro* by an incredible 38%. In 1994, the Solectria *Force* drove 11 miles (18 km) on one kilowatt-hour (kWh) of electricity, which costs on average (in America) 12 cents. So for the cost of one gallon of gas ($1.20), this car can travel 110 miles (177 km). This graph also shows...that heavier cars such as U.S. Electricar's mid-sized sedan or Ford Motor Company's van (*Ecostar*) will get fewer miles per kWh.

What is really exciting is that cars that are designed from the beginning to be electric vehicles, known as purpose-built EVs, are incredibly efficient...A lightweight, aerodynamic, two-person vehicle (Figure 14.10) built by MIT (Massachusetts Institute of Technology) demonstrated the ability of driving 26 miles (42 km) on 1 kWh of electricity, as did an electric motor scooter designed by the Schiller Group of Germany. When you think about the potential of combining a purpose-built EV with advanced energy storage systems, doubts about the performance of an EV vanish.

At the 1996 Northeast Sustainable Energy Association *Tour de Sol*, the purpose-built Solectria *Sunrise* commuter-type vehicle equipped with an Ovonic nickel-metal-hydride battery gave a range of 373.7 miles (601.3 km) at an efficiency of 11.24 miles (18.09 km) per kilowatt hour, as shown in Figure 14.14.

Note in Figures 14.12 and 14.14 the prominence of Ovonic's nickel-metal-hydride (NiMH) battery in vehicles with the greatest range. A recently released photo (Figure 14.15) shows Robert C. Stempel and Stanford R. Ovshinsky. Mr. Stempel is Chairman, Energy Conversion

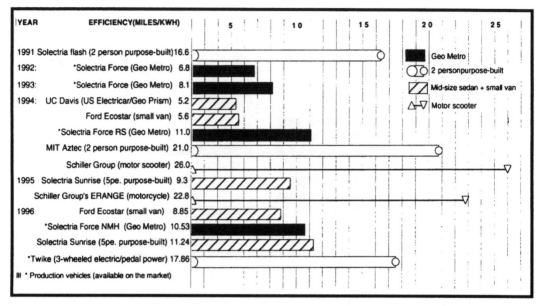

Figure 14.13. *Electric vehicle efficiency measured at the NESEA American* Tour de Sol *(1990 to 1996) (miles per kilowatt hour). (Nancy Hazard, Northeast Sustainable Energy Association)*

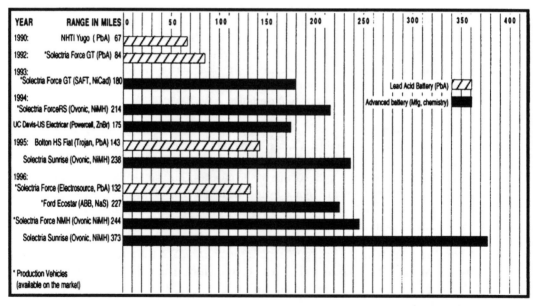

Figure 14.14. *Electric range measured at the NESEA American* Tour de Sol, *1990 to 1996 (miles driven between recharging). (Nancy Hazard, Northeast Sustainable Energy Association)*

*Figure 14.15.  Robert C. Stempel (left) and Stanford R. Ovshinsky (right), amid Ovonic batteries.  (Robert C. Stempel, 1997)*

Devices, Inc., and Chairman, Ovonic Battery Company, Inc.  Mr. Ovshinsky is President and CEO, Energy Conversion Devices, Inc., and CEO, Ovonic Battery Company, Inc.  This battery, "based on a proprietary, new fundamental approach to hydrogen storage alloys," is scheduled, it is said, to have a commercial release in 1998.  The price per kilowatt hour is, at this writing (1997), unknown.  The development of this new battery and the efforts of the two principals are well told by Michael Shenayerson in Ref. 6.

The next chapter presents the design status of solar-powered race cars in the late twentieth century.

# References

1. Wakefield, Ernest H., *History of the Electric Automobile: Battery-Only Powered Cars*, Society of Automotive Engineers, Warrendale, PA, 1994.
2. Pimental, David and Marcia, *Food, Energy & Society*, Arnold, London, 1979.

3.  White, David, General Manager, of Commercial Operations, Ayers Rock Resort, Yulara Northern Territory, Australia, personal communication.
4.  Hazard, Nancy, "Electric Solar Vehicle Race Opens New High Tech Opportunities for Women," Northeast Sustainable Energy Association news release, Greenfield, MA, 15 April 1992.
5.  Northeast Sustainable Energy Association, news release, 26 October 1994.
6.  Shenayerson, Michael, *The Car That Could,* Random House, NY, 1996.

# CHAPTER 15

# Solar Race Car Design: The Late Twentieth Century

## Changes in Design Standards

This chapter illustrates the status of solar-powered race cars toward the end of the twentieth century. The mutations have paralleled the modification in design exemplified by competitors in the much earlier established Indianapolis 500 race. Before World War I, the race car driver, who became the legendary Captain Edward Vernon (Eddie) Rickenbacker (1890–1973), drove boxy motorcars. The design of these vehicles varied substantially from one vehicle to the next.[a] Today, all Indianapolis 500 cars have, to a first approximation, identical shapes and similar drive systems. The same is true with solar-powered race cars. As Chapter 12 clearly illustrates, in fewer than a dozen years, the design of solar cars changed greatly under the pressure of race competition, in both the aerodynamic shapes and the drive systems.

---

[a] With America's entry into World War I, Rickenbacker, an early race car driver, was a chauffeur to General John J. Pershing, Commander of the U.S. Expeditionary Force in France. Later, joining a flying squadron, Rickenbacker became America's first Ace and was awarded the Congressional Medal of Honor and the *Croix de Guerre*. After discharge, Rickenbacker returned to the automobile industry, first with his own company and then with the Cadillac Motor Car Company. Subsequently (in 1938), Rickenbacker became president, heart, and soul of Eastern Airlines.

Supporting this statement, I received the following message from Dr. Fredy Sidler, director of the Ingenieurschule Biel in Biel, Switzerland, and leader of the winning team for the 1990 World Solar Challenge Race. This communication exemplifies the high standards being established for solar car races.

**Exclusive for the *Spirit of Biel II*: The Most Efficient Silicon Solar Generator That Was Ever Built.**

In 1985, the *Silver Arrow* of Mercedes-Benz, and one year later the *Sprit of Biel I* and the MEV-vehicle, occupied the winner's place in the Swiss *Tour de Sol* in different categories. These vehicles drove with solar generators manufactured by AEG (Germany). Since 1989, the solar activities in the Daimler-Benz group have been realized by Telefunken Systemtechnik GmbH, a subsidiary of Deutsche Aerospace.

The *Sprit of Biel II* of the engineering school Biel, which started in the World Solar Challenge (Race in) Australia, will enter the (1990) race with a silicon solar generator by Telefunken Systemtechnik. With this innovative generator, the company implements an additional large step forward into the future of highly efficient, photovoltaic energy transformation. In the preparatory phase for the project *Spirit of Biel II,* the Biel engineering school searched worldwide for a suitable manufacturer of a high-performance generator. The Biel engineering students approached the company with headquarters in Wedel near Hamburg via the general agency of Telefunken Systemtechnik in Switzerland. The challenge for the Wedel engineers consisted in developing a highly efficient solar generator that would guarantee an optimum of efficiency from the specified surface of the vehicle and would withstand the extreme demands of a rally over some 3000 km (1864 miles) across the (Australian) desert. At the same time, it was important to keep the weight (of the vehicle) as low as possible and to achieve a high resistance to fracture. And finally, the solar generator should be immune to the extreme climate of the desert. This catalogue of requirements was met by developing a new solar cell and a special generator technology.

The solar cells for this unique generator are manufactured from high-quality silicon basic material. In this context, the individual process steps are tailored for the minute interior electrical losses. This new type of cell for the *Sprit of Biel II,* instead of the traditional contact systems which cast shadows onto part of the light-sensitive front, is fitted with metal contacts 'standing' in some 100 hair-fine laser grooves. For all practical purposes, this overshadowing is eliminated, and the contacts are provided with an enhanced electrical conductivity.

As a result of the shingle-type arrangement of the individual solar cells, the Wedel specialists achieved a further reduction of electrical resistance and a packing density of 97.5% in the generator and thus a considerable increase of output surface. By encapsulating the solar cells in glass-fiber reinforced, highly transparent plastics, a generator that is merely 1.3 mm (0.5 in) thin was produced which, with a surface of almost 8 m$^2$ (86 ft$^2$), weighs 17 kg (37 lb), thus doing justice to the lightweight type of construction of the *Spirit of Biel II.* At the same time, the solar generator represents the mechanically stiffening surface structure of the vehicle. Generally speaking, solar generators today produce 100 to 120 watts per square meter. Equivalent peak values of space solar generators produce up to 140 watts.

In comparison, the solar generator built by Systemtechnik for the *Spirit of Biel* produced a much higher output of 170 watts per square meter. This (achievement) is a value which, for all practical purposes, is unique worldwide.

Table 15.1 shows data for the solar generator of the *Spirit of Biel II*. Table 15.2 describes the technical characteristics of that vehicle.[1]

### TABLE 15.1
### SOLAR GENERATOR DATA OF THE *SPIRIT OF BIEL II*

| | |
|---|---|
| Output | 1.3 kW (25 C° [77°F] AM 1.5 spectrum) |
| Efficiency | 17% |
| Surface | 7.67 m² (82.5 ft²) |
| Degree of Cell Packing | 97.5% |
| Thickness | 1.3 mm (0.05 in.) |
| Weight of Generator | 17 kg (37.5 lb) |
| Manufacturer | Telefunken Systemtechnik |
| Swiss General Agency | Elektron AG, CH-8804 Au/Zurich |

### TABLE 15.2
### TECHNICAL CHARACTERISTICS OF THE *SPIRIT OF BIEL II*

**Car Body:**

:General Properties:
Monoposte composite body optimized to light weight and little air resistance, safe traveling behavior in the whole field of application taking into account different cross-wind conditions.

Construction Form:
Supporting fiber reinforced body structure, a shell of body on layer lamination, strengthening ribs in sandwich construction, laminated by hand.

Materials:
Epoxy resin reinforced by carbon and aramid fibers

Data Specifications:

| | |
|---|---|
| Weight without mechanic electronic and solar cells | 66 kg |
| Drag coefficient Cw | 0.13 |
| Face surface | 1.1 m² |
| Dimensions (length × width × height) | 5620 × 2000 × 1000 mm |

Manufacturer:
Bucher lightweight constructions, CH-8117 Fällanden

## TABLE 15.2 *(continued)*

**Mechanical Elements:** (self-developed by the School of Engineering)
Suspension:
Front: Double triangle transverse control linkage with hydropneumatic suspension (spring mounting)
Rear: Longitudinal oscillator in the center of the car with incorporated start up bucking eliminator

Tires:
Double tires: racing bike tires 26-in., 19 mm wide; single tire: Bike-Slick 26 × 1.25 in.

Brakes:
Front:   Hydraulic disc brake
Rear:   Locking brake

Steering Assembly:
Steel cable steerage with guided middle part

Mode of Driving:
On rear wheel with chain

**Electric Elements:**
Solar Generator:
Construction: Highly efficient silicon solar cells in shingle technics (clap-board-technics), enclosed in ultra-lightweight glass-fiber reinforced plastic

Data Specifications:

| | |
|---|---|
| Output power (at 25°C [77°F] AM 1.5-Spectrum) | 1300 watts |
| Efficiency | 17% |
| Surface and thickness of generator | 7.67 m², 1.3 mm |
| Cell masking degree | 97.5% |
| Generator weight | 17 kg |

Manufacturer: Telefunken Systemtechnik

Maximum Power Tracker:
Up converter, self developed by the School of Engineering

| | |
|---|---|
| Nominal power | 220 watts (seven converters built in) |
| Efficiency at nominal power | 98.6% and 30°C (6°F) cooling medium temperature |
| Efficiency at 1/20 nominal power | 93% |
| Weight of a power tracker | 0.4 kg |

**TABLE 15.2** *(continued)*

Battery:
Silver-zinc battery, 86 cells at 1.5 V in series

| | |
|---|---|
| Nominal (rated) voltage | 129 V |
| Capacity appropriate of definition | 25 Ah (measured at C5 = 35 Ah) |
| Total weight of battery | 38 kg |

Manufacturer: Eagle-Picher USA

Motor:
Synchro-motor with magneto-permanent, self developed by the School of Engineering

| | |
|---|---|
| Nominal power | 1100 watts (peak power = 5000 watts) |
| Nominal voltage | ca. 65 V |
| Efficiency at rated load | 94.5% |
| Weight | 4.2 kg |

Electronics for Drive Units:
Inverter with MOS–FET, high efficiency and control system for run with cos $\varphi = 1$, self developed by the School of Engineering

| | |
|---|---|
| Nominal power | 1100 watts (peak power = 7000 watts) |
| Efficiency | 97–98% |
| Weight | 5.1 kg |

Instrument Panel Controls:
Battery voltage
Ampere-hour-counter charging and discharging control of the battery
Current in each of the seven power trackers
Tachometer
Power of the solar generator
Power of electronics
Powermat: Provides constant electrical power while driving

Lighting:
Stop lights, direction indicator, signal flasher (hazard warning) alimented by a separate 12-V silver-zinc battery

Weight:

| | |
|---|---|
| Empty weight | 175 kg |
| Total weight, including driver | 255 kg |

Source: Dr. Fredy Sidler, Ingenieurschule Biel, Biel, Switzerland, Ref. 1.

# Australian World Solar Challenge Race—1993

The World Solar Challenge Race of 1993 was the third in a series of this severe race. Apparent to independent observers is the improvement over time in the care of building and the performance of all vehicles. Similar to the more familiar Indianapolis 500 race held every Memorial Day in Indianapolis, Indiana, where competing cars are superbly designed and double-digit millions of dollars are required to field a car and team, the World Solar Challenge Race likewise has become expensive. As stated later in this chapter, the Honda *Dream* is believed to have cost $3 million, and the Ingenieurschule Biel *Spirit of Biel III* cost approximately $2.1 million. In the same way as in the Indianapolis 500 race, the vehicles are becoming more look-alikes. Compare the substantial vehicle structural differences in the 1987 World Solar Challenge Race as pictured in Chapter 13 with the first- and second-place finishers in the 1993 race (Figures 15.1 and 15.2). Likewise, excellent summaries of car characteristics have been written for these races.[2,3]

Table 15.3 shows the results of the 1993 World Solar Challenge Race.[4]

**TABLE 15.3**
**RESULTS OF THE 1993 WORLD SOLAR CHALLENGE RACE**

| Position | Team | Time (Hr:Min) |
|:--------:|------|:-------------:|
| 1 | Honda Research and Development | 35:28 |
| 2 | Ingenieurschule Biel | 38:30 |
| 3 | Kyocera Corporation | 42.35 |
| 4 | Waseda University | 42:50 |
| 5 | Aurora Vehicle Association | 43:00 |
| 6 | Toyota Motor Corporation | 46:34 |
| 7 | Northern Territory University | 46:50 |
| 8 | California Polytechnic University, Pomona | 47:21 |
| 9 | George Washington University | 47:46 |
| 10 | Zero to Darwin Project | 48:38 |
| 11 | University of Michigan | 49:07 |
| 12 | Nissan Motor Company | 50:21 |
| 13 | California State University, Los Angeles | 50:37 |
| 14 | Stanford University | 51:38 |
| 15 | Team Philips Solar Kiwi | 60:36 |
| 16 | Mabuchi Motor Corporation | 60:57 |
| 17 | Team Sofix | 64:56 |
| 18 | Tokia University | 74:22 |
| 19 | Monash University | 74:50 |
| 20 | Laughing Sun Racing | 75:48 |
| 21 | Mino Family Team | 76:21 |
| 22 | University of Oklahoma | 79:37 |
| 23 | Sonderborg Teknikum | 79:43 |

**TABLE 15.3** *(continued)*

| Position | Team | Time (Hr:Min) |
|---|---|---|
| 24 | Ashiya University | 79:48 |
| 25 | Dripstone High School | 81:17 |
| 26 | Panda-san | 84:15 |
| 27 | Team Solar Flair | 84:57 |
| 28 | KIA Motors | 85:27 |
| 29 | Team Alarus | 86:42 |
| 30 | Annesley College | 87:35 |
| 31 | Hokuriku Electric Power Company | 89:47 |

The following teams did not complete the race:

| Team |
|---|
| Hokkaido Automotive Engineering College |
| Team Doraemon |
| Solar Japan |
| Mitcham Girls High School |
| Morphett Vale High School |
| Team New England |
| Le Soleil |
| University of Western Ontario |
| Team TR 50 |
| NT Institute of TAFE |
| Meadowbank TAFE |
| Villanova University |
| JCJC Solar Car Club |
| The Banana Enterprise |
| Hama Yumeka Team |
| Team Heliox |
| Helio Det Team |
| Team Moscow |
| Team Holy Cheat |

Note that the time of finishing is more compact, indicating greater concern with reliability compared with the 1987 race as discussed in Chapter 13.
Source: David H. Goodman, Ref. 4.

Through the kindness of Dr. Fredy Sidler, director of Ingenieurschule Biel in Biel, Switzerland, summaries of the first- and second-place cars are presented in this chapter. Comments are given on the Honda *Dream* and subsequently the Swiss Ingenieurschule Biel *Spirit of Biel III*. This order of discussion follows the listing of these two cars at the finish of the race.

# The Honda *Dream*

Before comparing the first- and second-place winners in the 1993 World Solar Challenge Race, it is convenient to compare the General Motors approach to the 1987 race, which was the first race held in Australia. In designing the GM *Sunraycer,* the world's largest corporation set out with the firm principle of using the best talent worldwide to develop a solar-powered car that, barring unforeseen circumstances, would be a winner. Possibly GM had been stung by the claim that many commentators believed Japanese internal combustion cars in 1987 were more advanced in quality than American cars at that time. To compete in the World Solar Challenge Race of 1987, GM wisely engaged Dr. Paul MacCready as project manager. MacCready was one of the leading architects and builders of lightweight airborne vehicles.

Had not MacCready brought forth the first man-powered airplane to win the $100,000 Kremer Prize in 1977? Had not MacCready and his team designed and built the first solar-powered airplane to fly nonstop across the English Channel, winning the $200,000 Kremer Prize in 1979? Before the Australian World Solar Challenge Race was conceived by Hans Tholstrup, MacCready was already internationally known. Matching the man and his team at his company, AeroVironment Inc., was emphasis by GM to support MacCready's organization with the best the world had to offer in knowledge and accessories. In addition, although the budget made available was never divulged, it was believed huge at the time. Furthermore, the two vehicles constructed were thoroughly tested, and the whole was cemented together as a highly skilled team led by veteran Howard G. Wilson, vice president of Hughes Aircraft (retired), which was then a GM subsidiary.

The result was that the GM *Sunraycer* was a clear winner over the second-place car in the 1987 race by almost a day. In showing to the world its competence, General Motors transferred from competition to promotion and sponsorship of solar car races for colleges and universities in North America. Hence, the GM Sunrayce 90–USA was born. Discussion of that race is omitted here because of space limitations.

It may be surmised that the Honda Corporation, a non-entry in the 1987 race, would play the game of persistence in planning for the series of World Solar Challenge Races. Honda placed second in 1990 and first in 1993 with its *Dream* (Figure 15.1).

The Honda *Dream* is a beautiful and well-designed three-wheeled vehicle. Its length is slightly less than 6 m (20 ft), its width is 2.0 m (6.5 ft), and its height is 1.02 m (3.35 ft). The drag coefficient is a low 0.10, and the frontal area is 1.14 $m^2$ (12 $ft^2$). The chassis and body are based on carbon fiber in honeycomb. The motor is of permanent magnet, synchronous type, rated at 1.5 kW continuous rated power, with 95% efficiency. It weighs 12.8 kg (28 lb). There are 83 silver-zinc batteries of approximately 5 kWh capacity. Silicon solar cells of 21.2% efficiency complete the drive system. Without the driver, the vehicle weighs 187.5 kg (413 lb). Drivers of the car have reported *Dream* accelerates rapidly; at high speed, "the body shook unmercifully and reverberated with rattling sounds and road noise."

Table 15.4 provides data on the Honda *Dream.*

*Figure 15.1. The 1993 solar-powered Honda* Dream. *(Dr. Chester R. Kyle)*

## TABLE 15.4
## DATA SHEET FOR THE HONDA *DREAM*

| | |
|---|---|
| **Place:** | First outright |
| **Class:** | First Silicon/Silver-Zinc Class |
| **Car:** | 2, *Dream*, Honda R&D |
| **Country:** | Japan |
| **Team Leader:** | Takahiro Iwata |
| **Address:** | Honda R&D Ltd., Wako Center, 1-4-1 Chuo, Wako-Shi, Saitama, 351-01 Japan |
| **Chief Engineer:** | Takahiro Iwata |
| **Drivers:** | Masashi Yamamoto, Jo Yoshida, Toshihiro Miki, Takahiro Iwata |
| **Team Members:** | 22 team members |
| **Sponsor:** | Honda R&D |
| **Cost:** | US$1,500,000 (1 car) |
| **Project Time:** | One year |

**Performance:**

| | | | |
|---|---|---|---|
| Qualifying speed | 125.0 km/h | Qualifying position | Second |
| Est. top speed (solar/battery) | 92/130 km/h | Predicted race avg. | 86.0 km/h |
| Actual average speed | 84.96 km/h | Best day avg. speed | 90.8 km/h |
| Predicted distance/day | 774 km | Actual avg. distance/day | 747.5 km |
| Best distance/day | 802 km | Practice distance | 3000 km |
| Time for race (hr:min) | 35:28 | Penalties | None |

**Daily Results:**

| Day | 1 | 2 | 3 | 4 | 5 |
|---|---|---|---|---|---|
| Distance (km) | 746 | 746 | 807 | 691 | 23 |

**Aerodynamics:**

| | | | |
|---|---|---|---|
| Drag coefficient ($C_d$) | 0.10 | Frontal area (A) | 1.14 m² |
| Drag area ($C_dA$) | 0.114 m² | Measured | Full-size wind tunnel and power measurement on track |

## TABLE 15.4 *(continued)*

**Dimensions:**

| | | | | | | |
|---|---|---|---|---|---|---|
| Weight (without driver) | 187.5 kg | | | | | |
| Length | 5.975 m | Width | 2.000 m | Height | | 1.020 m |
| Wheelbase | 2.250 m | Track | 1.340 m | Ground clearance | | 0.145 m |

**Wheels and Tires:**  Wheels are six-spoke, machined magnesium alloy covered with carbon fiber disc covers. Tires are IRC, front is 22 × 1.75, rear is 22 × 2.00 in. at 690 kPa.

| | | |
|---|---|---|
| Number | 3 | |
| Number of flat tires during race | 6 | |
| Rolling resistance (measured) | 0.0040 | |
| Time to change | Front: 2 min | Rear: 1.50 min |

**Brakes, Suspension, Steering:**  Front brakes are hand-operated hydraulically activated discs, rear is regenerative. Front suspension Is magnesium arm double wishbone, with coil over shocks. Rear is a magnesium mono swing arm, with coil over shock. Steering is by bar handle with wire link between Pittman arms.

**Chassis/Body:**  Chassis is a three-tub based carbon-fiber reinforced aramid honeycomb. Body is similar construction, one piece with canopy.

**Motor:**  Honda designed d-c brushless wheel/hub motor. Neodymium magnets, 6.0 kW maximum rated power, 1.5 kW continuous rated power, 1000 rpm (corresponds to 130 km/h), 95% efficiency, 12.8 kg.

**Controller:**  Honda PWM.

**Transmission:**  None, as wheel motor.

**Controls:**  Hand-operated throttle, hand-operated brake (controls similar to scooter). Constant speed or power cruise control. Full instrumentation.

**Telemetry:**  Full telemetry (70 bytes every 2 seconds) and voice radio.

**Batteries:**  Yuasa AgZn, 83 cells, 4.980 kWh capacity, 124.5 V, 40 Ah rated, 62.3 kg. Charge at 17:00 each day (%): 43, 60, 49, 14%.

**Solar Cells:**  Sun Power USA monocrystalline silicon, 21.2% cell efficiency.

**Type of Solar Panel:**  Curved, fixed, single top panel split into 10 facets, 4682 cells in 44 major panels, cell weight 3.01 kg. Cells are encapsulated with a front surface of textured sheet designed to enhance energy collection. Total area 8.278 m², 124 V panel voltage.

| | |
|---|---|
| **Maximum Instantaneous Panel Power During Race:** | 1800 W |
| **Typical Maximum Panel Power at Midday During Race:** | 1520 W |
| **Average Panel Power from 8:00 A.M. until 5:00 P.M.:** | 1232 W sunny, 948 W average |
| **Speed on 1000 W to Motor:** | 80 km/h |
| **Tracker:** | 10 Honda-designed peak power point trackers, 98% efficient, 3.5 kg weight |

**Notes and Problems During Race:**  Day 1: For first 250 km of race, lack of power to motor. Stopped at 250 km because of rear tire puncture and changed controller and wheel/motor/tire in 7 minutes.
Day 2: Changed front left and rear tires during afternoon.
Day 3: One puncture.
Day 4: Total loss of power on a hill. Stopped, cycled main power switch. Thought to be a computer crash/hang; lost 3.5 minutes. Rear flat, 2 minutes. Front left puncture, both front tires changes; 4 minutes lost.

Source:  Chester R. Kyle, Ref. 2.

By reading that description and comparing the characteristics of both the Swiss *Spirit of Biel II* and the Honda *Dream,* you can appreciate the high level of development the best solar-powered cars have reached. The latest cars, such as the Indianapolis 500 race cars, are now almost indistinguishable in physical shape. Contrast that near unanimity in design with earlier solar-powered cars.

## Ingenieurschule Biel *Spirit of Biel III*

After reading Markus Liniger's detailed report[3] on the Ingenieurschule Biel highly refined *Spirit of Biel III* (shown in Figure 15.2), I asked Dr. Sidler for his analysis of why the Honda *Dream* arrived at the finish line three hours, two minutes earlier than the *Spirit of Biel III.* Dr. Sidler's reply, written in his letter of 1 May 1995,[5] was as follows:

> There might be several reasons (why the Honda *Dream* crossed the finish line before the *Spirit of Biel III*). One of them was our problem with the wheel cover touching the tire and slowing down the speed of our car. The problem started at Katherine (Northern Territory, Australia) at noon of the first day, three hours after the start. We were about 20 minutes in front of Honda. We fixed the problem only at the end of Day Two, close to Alice Springs where Honda was 60 km (37 miles) in front of us. This (fault) was not decisive for the race.

> Analyzing the whole situation, I would say that the total efficiency of our drive system (from the peak power trackers to the wheel) of 96.5% was—I think—a little better than Honda's. The air resistance of the two cars was—according to my information—equivalent. The rolling resistance of our car must have been better because of the weight (*Dream*

*Figure 15.2. The* Spirit of Biel III. *(Dr. Fredy Sidler)*

was about 30 kg [66 lb] heavier than the *Spirit of Biel III*) and because of the adhesive friction of our Michelin tires (f = 0.0055.) To be honest, I doubt the Honda adhesive friction coefficient of f = 0.0040. It seems that there were certain differences among almost all teams about the method of measuring of this value.

The reason why we really lost the race is the solar cells. We had a total panel efficiency of 18.8% (measured during the race). Honda's efficiency was 21.6%. (This figure is confirmed by T. Iwata, the Honda team manager.) In short: Honda had nearly 15% more energy than we had. Our cells were produced by Deutsche Aerospace (a German company); Honda's cells were produced by the California company Sun Power (Dick Swanson).

# The Motor-Wheel of the Swiss *Spirit of Biel III*

Aerodynamically, toward the end of the twentieth century, the teardrop design of the General Motors *Sunraycer* had triumphed. Batteries and motors were treated in *History of the Electric Automobile: Battery-Only Powered Cars.*[6] However, what has happened with transmitting torque to the wheel, an important factor in reaching maximum efficiency for such a vehicle? Markus Liniger of Ingenieurschule Biel in Biel, Switzerland, has provided an elegant solution,[3] first postulated and employed (with a d-c motor) by Dr. Ferdinand Porsche while employed by Lohner, coach-builder for the Austrian-Hungarian court, and shown at the Paris Salon in 1900.

At the time, Porsche said[7]:

> The trouble with most horseless carriages is the complexity of their transmission. All this business of shafting, bevel gears, and chains could be quite simply avoided if we used electric cables to carry the power to the place where it is needed, which, obviously, is at the driving wheel.

Of Engineer Porsche's motor, Liniger writes (shortened and edited):

> We developed a so-called hubbed-wheel motor, which means that the motor forms the hub of the rear wheel. From the electrical point of view, we have a d-c brushless motor and an electronically commutated d-c motor (brushless) with permanent magnets (a three-phase synchronous motor), which excels at the highest rate of efficiency. The engine was developed, built, and measured at the Biel School. Thanks to its position in the rear wheel, mechanical losses linked with transmission from engine to wheel (chain) are avoided. Consequently, the engine must run more slowly and produce high torque, as no gear-shifting is possible anymore. Because in an electrical motor the torque is proportional to the volume, a hubbed-wheel motor is necessarily heavier than a faster-running motor with the same power. In addition, there are problems with the unsprung mass. Thanks to the use of special materials, a very good ratio between torque and mass at the highest level of efficiency was achieved. Special attention has been paid to the designing of a sturdy enough construction to protect the motor from shocks that can only be attenuated by tires...

<u>Mechanical Structure</u>: (Figure 15.3)...shows a section of a motor. A steel, hollow shaft, the end of which fits directly into the suspension system of the rear wheel, carries the whole motor. In the middle of the shaft is located the stator. The rotor consists of a ring. A flange is screwed on both sides. The rotor is fixed with two steel, corrugated ball-bearings on the shaft. Two lids close the bearing and support the axial forces from the rotor onto the shaft...(Figure 15.4) displays the shaft with the stator. The power and control cables are led through the hollow shaft...The iron package stator is fixed onto the stator carrier...The materials used produce very low eddy current and hysteresis losses.

<u>Winding</u>: The winding is stored in the grooves of the 28-pole motor. The spools, each with two loops, are connected in serial. All three phases are connected as star or wye...

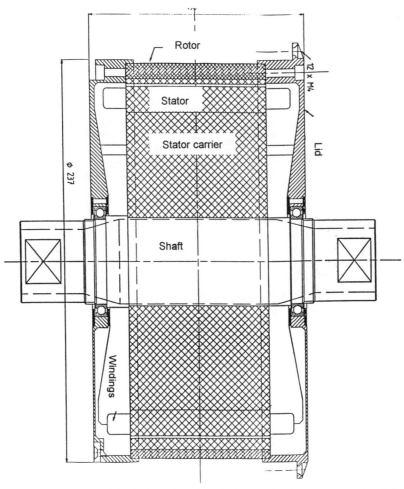

*Figure 15.3. Section of the* Spirit of Biel III *hubbed-wheel motor. (Dr. Fredy Sidler)*

*Figure 15.4. Support of the stator (top) and the stator itself (bottom) for the* Spirit of Biel III. *(Dr. Fredy Sidler)*

Technical data of the motor are as follows. (All figures are nominal. If designing a vehicle, read Liniger's paper cited in Ref. 3.)

| | |
|---|---|
| Weight | 12 kg |
| Weight of whole rear wheel | 14.1 kg |
| Outside diameter of motor | 237 mm |
| Voltage | 110 volts |
| Phase current | 12.5 amperes |
| Variable frequency and pulse-width modulation | 207.5 Hz |
| RPM | 888 1/min |
| Speed of vehicle | 89 km/h |
| Torque | 14.5 Nm |
| Power | 1350 W |
| Efficiency | 96% |

Table 15.5 provides more data on the *Spirit of Biel III*.

## TABLE 15.5
## DATA SHEET FOR THE *SPIRIT OF BIEL III*

| | |
|---|---|
| **Place:** | Second outright |
| **Class:** | Second Silicon/Silver-Zinc Class |
| **Car:** | 1, *Spirit of Biel/Bienne III*, Engineering College of Biel |
| **Country:** | Switzerland |
| **Team Leader:** | Fredy Sidler |
| **Chief Engineer:** | René Jeanneret |
| **Address:** | Ingenieurschule HTL, Quellgasse 21, Postfach 1180, 2501, Biel, Switzerland |
| **Drivers:** | Paul Balmer, Pascal Felder, Hans Grünig, Jean-Michel Molinari |
| **Team Members:** | P. Baumann, P. Bucher, M. Bühler, R. Christe, H. Gochermann, J. Graells, W. Kraehenbueler, M. Laminet, A. Linder, M. Liniger, E. Ruppert, M. Wittwer |
| **Sponsors:** | Swatch, Swiss Engineering, Swiss Government, City of Biel-Bienne, Flugzeugwerke Emmen, Michelin, Vacuumschmelzze/Hanau |
| **Cost:** | US$2,100,000 |
| **Project Time:** | Two years |

**Performance:**

| | | | |
|---|---|---|---|
| Qualifying speed | 129.9 km/h | Qualifying position | First |
| Est. top speed (solar/battery) | 90/145 km/h | Predicted race avg. | 88.0 km/h |
| Actual average speed | 78.27 km/h | Best day avg. speed | 85.0 km/h |
| Predicted distance/day | 792 km | Actual avg. distance/day | 675.8 km |
| Best distance/day | 737 km | Practice distance | Many km |
| Time for race (hr:min) | 38:30 | Penalties | None |

**Daily Results:**

| Day | 1 | 2 | 3 | 4 | 5 |
|---|---|---|---|---|---|
| Distance (km) | 723 | 703 | 737 | 540 | 310 |

**Aerodynamics:**

| | | | |
|---|---|---|---|
| Drag coefficient ($C_d$) | 0.10 | Frontal area (A) | 1.05 m² |
| Drag area ($C_dA$) | 0.105 m² | Measured | Power measurements on Michelin track |

## TABLE 15.5 *(continued)*

**Dimensions:**

| | | | | | | |
|---|---|---|---|---|---|---|
| Weight (without driver) | 157.5 kg | | | | | |
| Length | 5.700 m | Width | 1.998 m | Height | 1.000 m |
| Wheelbase | 2.400 m | Track | 1.500 m | Ground clearance | 0.243 m |

**Wheels and Tires:**  J16-MT 1.85 carbon fiber disc, 515 × 70 mm (cross section); Michelin radial, 65/80-16 at 500 kPa.

| | |
|---|---|
| Number | 3 |
| Number of flat tires during race | 1 |
| Rolling resistance (measured) | 0.0055 |
| Time to change | Front: 3 min    Rear: 5 min |

**Brakes, Suspension, Steering:**  Hydraulic discs on front, mechanical hand-operated rim calliper on rear, no regenerative braking. Front suspension is double wishbone, with elastomer shock/spring. Rear is carbon fiber trailing arm, integral hubbed-wheel motor. Steering is joystick activated steel cables, with redundant cables and transverse control link.

**Chassis/Body:**  Chassis is a carbon honeycomb sandwich. Body is a carbon fiber/Nomex honeycomb epoxy composite.

**Motor:**  Biel (Jeanneret designed) wheel motor, d-c brushless, 1.4 kW continuous power rating, 11.0 kW maximum power rating, 125 V, 900–1500 rpm, convectively cooled, 12 kg with control electronics, 97.5% efficient at 1.4 kW.

**Controller:**  Biel designed, 98.5% efficient MOSFET inverter, torque regulation with Tempomat and Powermat, 98+% efficient, 5 kg.

**Transmission:**  None, as wheel motor.

**Controls:**  Foot speed control and brake, hand brake, joystick steering.

**Telemetry:**  Voice radio link (165 MHz)—vehicle data relayed to support bus, where it was entered into a Mac Powerbook 180. Excel spreadsheet and Labview used for real-time analysis.

**Batteries:**  Eagle-Picher SZHR 25-5, 82 cells, 123V, 4.92 kWh capacity, 40 Ah rating, 39.4 kg. Charge at 08:00/17:00 each day (%): 100/29, 66.2/22, 67/13, 33/17, 42/9%.

**Solar Cells:**  Deutsche Aerospace, monocrystalline silicon, Hi-Eta (BSFR), 19.0% individual cell efficiency.

**Type of Solar Panel:**  Single, fixed, very gently curved top surface, split into three facets. Total panel weight 9.8 kg. 3610 cells are double shingled (98.4% fill factor) and encapsulated in resin, making 16 main panels and 2 front panels. Total array voltage: 118 V. Peak power: 1460 W at standard solar illumination.

| | |
|---|---|
| **Maximum Instantaneous Panel Power During Race:** | 1750 W |
| **Typical Maximum Panel Power at Midday During Race:** | 1391 W |
| **Average Panel Power from 8:00 A.M. until 5:00 P.M.:** | 1117 W |
| **Speed on 1000 W to Motor:** | 74 km/h |

**Tracker:**  Three Biel-designed maximum power point trackers (up-converters), 99% efficient at 600 W, 98.6% at 900 W, 97.3% at 20 W. Total weight 2.7 kg.

**Notes and Problems During Race:**  Day 1: When leaving the Katherine media stop, hit a deep curb. Reduced speed by 14 km/h—left front wheel spat was rubbing against the tire.
Day 2: Left front wheel spat still rubbing. Removed late in the day. Also changed one peak power tracker.
Day 4: 3.5 minutes to change punctured front tire. Changed tires regularly at the end of each day.

Source: Chester R. Kyle, Ref. 2.

# A Commercial Market for Solar-Powered Cars

Does a commercial market exist for solarmobiles? Some *aficionados* believe a small one exists in the 1990s and beyond, in the $200,000 to $300,000 class if the solarmobiles are sold in the same way as sail yachts are marketed. Solarmobiles are similar to a yacht, which is a sports vehicle, and some believe that market is the one to be originally sought. A yacht requires water on which to sail; likewise, solar-powered vehicles need the sun. An entrepreneur wishing to assemble and offer solar vehicles has possibly the best chance of success in those areas of a country which are generously blessed with many sun-hours per year. Chapter 11 lists the percentage possible sun-hours of many American cities.

What makes solarmobiles commercially viable in the 1990s and beyond, when electric vehicles converted from commercial vehicles in the late 1960s and early 1970s were not? Solarmobiles have substantial range in daylight hours. More than 480 miles (770 km) have been logged in a day. The converted electric vehicles of the 1960s and 1970s were fast—60 mph (100 km/h)—but their range in city driving was limited to approximately 30 miles (50 km). In all but exceptional cases, that distance is too small; psychologically, it is absolutely too small. For instance, *Thunderbolt 240* (a Ford *Pinto* station wagon), the "240" representing the charging voltage, was pronounced by Commonwealth Edison Company (the northern Illinois electric utility) spokesman, William F. Shafer (retired) as the best of the conversions (at the time).[8] However, the vehicle was unable to travel from Evanston, the first suburb north of Chicago, to O'Hare International Airport and back on a single charge.[9] Even the well-conceived and revolutionary *Sundancer*, built as an electric vehicle from the ground up, had too little range with its lead-acid batteries.[10]

To overcome range deficiency, Dr. Victor Wouk reintroduced the concept of hybrid electric vehicles—motor cars with two sources of power. In Wouk's case, this meant a gasoline engine and batteries.[11] In reintroducing the concept of multi-powered sources for vehicles, Wouk paved the way for the solarmobile because they too are hybrid-powered vehicles—solar cells and batteries. The multi-power idea is old. Count Felix Carli, an ingenious Italian textile magnate, realized in 1894 the range limitation of electric vehicles and supplemented available power for his vehicle by placing in the drive train the potential energy of a wound spring.[12] Justus B. Entz, chief engineer of the Electric Storage Battery Company of Philadelphia, operated an electric vehicle augmented with a gasoline engine on 4 May 1897.[13] Both Entz's and Wouk's vehicles and the many other types of battery-only powered vehicles introduced,[6] while serving to initiate the concept of single and dual power sources, were too complex. Has the availability of solar cells changed the equation? Figure 15.5 shows a series of environmentally sensitive vehicles built at Dartmouth College under the supervision of Douglas A. Fraser. *Sunvox IV* climbed Mt. Washington, 6288 ft (1917 m). In discussing the commonness of a solar-powered car, Douglas A. Fraser said, "I want a solarmobile so I can jump in and drive to the post office."[14]

*Figure 15.5. Douglas A. Fraser's vehicles. Top, pictured left to right:* Sunvox IV, Sunvox I, *and* Ecovox *(ethanol fueled hybrid), and student team with trophies. Bottom,* Sunvox IV, *showing ingress. (Douglas A. Fraser)*

# Starting a Solarmobile Business—Traits of the Entrepreneur

Some say solarmobiles should be offered commercially in two styles: 1) body-plated solar cells, such as those in the GM *Sunraycer,* and 2) "pool-table" arrays, similar to those in the Ford *Sunchaser.* Chester R. Kyle, writing in 1987 and cited in Chapter 13, believes the optimum shape for solarmobiles is yet to be reached.[15] *Apropos* of vehicle shapes, the present configuration of today's automobile was finally determined only in 1902 by Walter C. Baker with his *Torpedo*, illustrated in Chapter 15 of *History of the Electric Vehicle: Battery-Only Powered Cars*,[6] 19 years after the first motorcar, an electric, in 1881.[16]

After founding a world-class nuclear instrument company, selling it, and retiring at age 47, I discovered, to build a successful business, certain commonalties that existed among the founders (who were alive) of 19 high-technology businesses that I analyzed. The following describes the results of my analysis.

Although variants exist from the so-called norm of these unique individuals, the founder of a high-technology company in America typically is a white male, born in a comparatively small community of relatively affluent parents. He had a Tom Sawyer type of early life, and he pursued creative jobs and earned money as a teenager. He attended a university, was a superior student, and was more likely to be interested in cultural activities than in sports. He became a teaching assistant and, on completion of formal academic training, had earned two science degrees. No founder of the 19 companies studied possessed a Master of Business Administration degree before initiating his company.

The founder-to-be was employed in two industrial jobs prior to starting his own business at age 31. He began his enterprise on very little capital typically furnished by himself, a close friend, or family. The Small Business Administration was neither consulted nor approached. A local bank initially was contacted, and credit was established.

The entrepreneur knew three essentials were necessary for success: The product or service must be wanted, that he had to be world class in knowledge of his offerings, and that high profit margins were a necessity. The founder designed his first product as a variant on others he had seen. The first offering was an immediate success. He also was the prime mover in subsequent company products and was always involved in technical discussions of new company creations. A regular duty of the founder was to maintain university or national-laboratory contacts. He was always interested in sales, in meeting with experts in his field, and in visiting with customers to ascertain their needs.

The typical founder appears to possess a remarkably stable family life. He married a college sweetheart at age 24, entered matrimony only once, and had no divorce. He and his spouse had four children in the period when this number was the norm. A long-term and excellent relationship existed between the founder and his children. The founder's hobbies are reading nonfiction and listening to classical music.

In later life, the founder monitors his company in which most of his assets are found. In addition, he is active in community affairs. Finally, the founder appears to have eminent health. Many founders have lived well beyond the allotted 'three score and ten' years. These achievers seem to be well-balanced individuals.

Regarding entrepreneurs in general, in contrast to the strictly high-tech company founders cited here, Arthur Levitt, Jr. and Jack Albertini have summarized the findings of McKinsey & Co.[17] Their report is based on a two-year study of members of the American Business Conference, a coalition of mid-sized, high-growth companies that have doubled in sales during 1975 to 1980. A successful entrepreneur is a consummate salesman,

> ...who radiates enormous, contagious self-confidence that often translates into respect and esteem from the employees...(They) are fanatics for fundamentals...(They) pay dogged attention to their firm's finances, operations, and external forces...Entrepreneur CEOs want to encourage experimentation...(They) spend more than 20% of their time on employee development, and one of their predominate traits is perseverance to the point of obsession...He is a fiercely independent, highly motivated, economic animal who shuns the herd instinct and prefers to match wits with the world on his own terms...(Finally), there is the appetite for daring.

Based on that description, are you this type of person? If so, then design, build, and start offering solarmobiles. Names of suppliers of all elements are scattered through this book. In point of development, solarmobiles of the 1990s, despite the sophistication of some, appear to be at a stage similar to where automobiles were in 1892, a period when E.P. Ingersoll, who later became editor of the 1895-established *The Horseless Age*, apocryphally wrote, "These vehicles were of a nature that only the inventor would dare ride in them."[18]

Solarmobiles also appear to be in the period of the airplane *circa* 1913, before its accelerated development during World War I. Anyone familiar with the many dead-end paths personal conveyances have followed will have little sanguinity to make a prediction for solar-powered personal vehicles. As previously mentioned, Douglas A. Fraser, chief designer of the Dartmouth College *Sunvox 1, II, and III* vehicles, expressed a desire to have a solarmobile that "I could jump into, much as I now do with my car, and be able to go to the post office with little concern."[14]

# The Sports Solarmobile

In America there are possibly more than 100,000 sail yachts, each valued at more than $200,000. That price range may be the market for the solarmobile if it is made of impeccable workmanship and quality, bearing small, appropriate, and warm elements of honed and varnished Stradivarius-style wood, with glove-leather cockpit accouterments to soften the austere, aseptic nature of some of the presently well-designed solarmobiles. For the possible recoil by future commercial builders to the previously mentioned price range for sun-powered cars, this is Rule One in starting a business: Large margins must exist to ensure that the inevitable business errors will not fatally destroy the enterprise. Another factor, surely to impact the use on public roads of the relatively fragile-structured solarmobiles, is the attitude of the transportation licensing authorities as demonstrated in Figure 15.6. This figure shows a letter received by Douglas A. Fraser from the State of New Hampshire when he applied for an operating license. From all of these reasons, some credence for considering the initial use of solarmobiles strictly as sports vehicles is warranted. More pragmatic and lower-priced sun-powered wheels may come later. If the history of personal-powered transportation can teach those designers willing to learn, the early steam, electric, and internal combustion engine vehicles followed this course.[6]

Finally, can we now write of the solarmobile as did E.P. Ingersoll in the November 1895 first issue of *The Horseless Age*[18]:

> At this formative period of the motor vehicle industry, exact data are difficult to obtain. All over the country, mechanics and inventors are wrestling with the problems of trackless traction. Much of their work is in an unfinished state; many of their theories lack demonstration. But enough has already been achieved to prove absolutely the practicability of the motor vehicle. What is here presented, however, is merely an earnest of what is to come.

I and possibly many of you await that entrepreneur.

Figure 15.7 is a rare photograph showing how the concept of a solarmobile has changed during an eight-year interval. The drive systems also have changed impressively. These identified vehicles are products of Ingenieurschule Biel, Höhere Technische Lehranstalt des Staates Bern (HTL), in Switzerland. Director of the engineering school, Dr. Fredy Sidler, kindly furnished pictures of this assembly. Personnel of Ingenieurschule Biel and their vehicles have participated in solar car races since 1985. In the Australian World Solar Challenge Races, their entries have placed third in 1987, first in 1990, and second in 1993.

# Influence of *Sunraycer, Spirit of Biel III,* and *Dream* on Design Today

Two high-level solar-powered race cars were fabricated by students from Massachusetts Institute of Technology (MIT) and from the University of Minnesota for Sunrayce 95. Notice how closely the two vehicles aerodynamically resemble the Honda *Dream*, the *Spirit of Biel III,* and the General Motors 1987 *Sunraycer*, whose scientific design has been detailed.[19] Figures 15.8 and 15.9 show the two university cars, respectively.

Unique for the cited MIT and Minnesota solarmobiles was their first- and second-place finishes with an elapsed time difference of only nine seconds over an 1850-km (1150-mile) course from Indianapolis, Indiana, to Golden, Colorado.[20] Moreover, in its 1995 initiated rules for the Sunrayces, the U.S. Department of Energy has wisely specified low-cost lead-acid batteries and economical, terrestrial-type solar cells, thus eliminating the advantages previously enjoyed by more wealthy and larger colleges and universities. According to Paul A. Basore of the

603/271-2484

*The State of New Hampshire*
*Department of Safety*
*James H. Hayes Building — Hazen Drive*
*Concord, N.H. 03305*

RICHARD M. FLYNN
COMMISSIONER OF SAFETY

ROBERT K. TURNER
DIRECTOR
DIVISION OF MOTOR VEHICLES

May 27,1988

Mr. Leonard Greenhalgh
Professor of Management
Dartmouth College
Hanover, New Hampshire  03755

Dear Professor Greenhalgh:

Please be advised that after investigation by personnel of the Department of Safety, it has been decided to allow Dartmouth College to register the experimental solar powered vehicle for use on public ways.

The sole intent of this approval is to allow the vehicle to be tested in normal road use incidental to research on the vehicle - <u>not general use, or interstate highway use.</u>

Due to irregular construction and components, we will also waive the need to heve the vehicle inspected by an official Inspection Station as meeting conventional inspection requirements.  However, we must mandate that at a minimum, the construction and equipment of the vehicle be no less than stated in Highway Enforcement Officer Strickford's report that is attached to this letter of approval.

I would recommend that a copy of this letter be carried in the vehicle along with the registration as required by law.

Sincerely,

Robert K. Turner
Director of Motor Vehicles

RKT/vlc

*Figure 15.6.  Permissive letter from the State of New Hampshire.  (Douglas A. Fraser)*

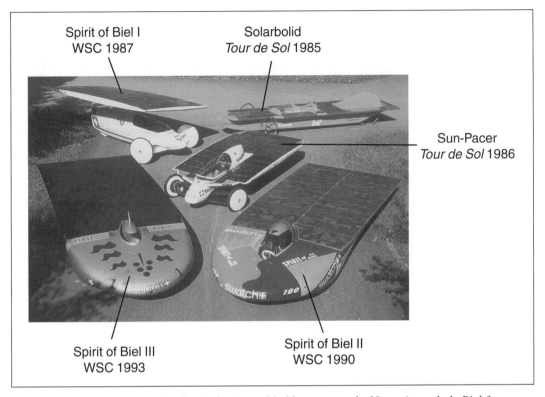

*Figure 15.7. Solar-powered vehicles assembled by personnel of Ingenieurschule Biel from 1985 to 1993. (Dr. Fredy Sidler)*

*Figure 15.8. The Sunrayce 95 winner, MIT #17. (U.S. Department of Energy, Byron Stafford. Photo by Warren Gretz, NREL)*

*Figure 15.9. The Sunrayce 95 second-place winner, University of Minnesota #35. (U.S. Department of Energy, Byron Stafford. Photo by Warren Gretz, NREL)*

U.S. Department of Energy, "the winning team this year (1995) spent only US$60,000 on their effort, and another top-ten team spent just US$20,000." In contrast, in Sunrayce 93, the University of Michigan *Maize & Blue* cost more than US$900,000.

The next chapter outlines some of my thoughts, based on my experience of more than 30 years with battery and hybrid electric cars.

# References

1. Sidler, Fredy, Director, Ingenieurschule Biel, personal communication, Biel, Switzerland, 5 April 1995.
2. Kyle, Chester R., *et al., Solar Racing Cars: 1993 World Solar Challenge,* Australian Government Publishing Service, Canberra, 1994.
3. Liniger, Markus, *A Trip Through Australia with the Spirit of Biel III,* Ingenieurschule Biel, Biel, Switzerland, 1994.
4. Goodman, David H., *Team Handbook*, University of Michigan, Ann Arbor, MI, 1994, pp. 44–45.
5. Sidler, Fredy, Director, Ingenieurschule Biel, personal communication, Biel, Switzerland, 1 May 1995.
6. Wakefield, Ernest H., *History of the Electric Automobile: Battery-Only Powered Cars*, Society of Automotive Engineers, Warrendale, PA, 1994.
7. Barker, Ronald, and Harding, Anthony, eds., *Automobile Design: Twelve Great Designers and Their Work*, Society of Automotive Engineers, Warrendale, PA, 1992.

8. Shafer, William F., personal communication, 1970.

9. Wakefield, Ernest H., president of Linear Alpha, Inc., the builder, was the driver.

10. McKee, Robert S., Borisoff, Boris, Lawn, Frank, and Norberg, James F., "Sundancer: A Test Bed Electric Vehicle," SAE Paper No. 720118, Society of Automotive Engineers, Warrendale, PA, 1972.

11. Wouk, Victor, "An Experimental ICE/Battery-Electric Hybrid with Low Emissions and Low Fuel Consumption Capability," SAE Paper No. 760123, Society of Automotive Engineers, Warrendale, PA, 1976.

12. *Scientific American,* **LXX,** 16, 21 April 1894, p. 251.

13. Maxim, Hiram Percy, *Horseless Carriage Days,* 2nd ed., Harper, New York, 1937, pp. 144–149.

14. Fraser, Douglas A., Dartmouth College, personal communication, 2 November 1988.

15. Fries, Peter O., "The 1987 World Solar Challenge: A Technical Report for the New Energy Development Organization of Japan," Renewable Energy Services, P.O. Box 33, St. Lucia, Queensland, 4067 Australia; report prepared for Energy Promotions, 1697 Pittwater Rd., Mona Vale, NSW, (a division of Flipper Scow Pty. Ltd.), with funding provided by New Energy Development Organization, January 1988.

16. Barral, Georges, *Histoire d'un Inventeur, Gustave Trouvé,* Carre, Georges, ed., Paris, 1891, p. 610.

17. Levitt, Arthur, Jr., and Albertini, Jack, "The Successful Entrepreneur: A Personality Profile," *The Wall Street Journal,* 29 August 1983, p. 12.

18. Ingersoll, E.P., Salutory, *The Horseless Age*, Vol. 1, No. 1, November 1895, p. 1.

19. MacCready, Paul, *GM Sunraycer Case History,* Society of Automotive Engineers, Warrendale PA, 1992.

20. Basore, Paul A., *Sunrayce 95, Progress in Photovoltaics: Research and Applications*, Vol. 3, John Wiley & Sons, New York, 1995, pp. 425-427.

# CHAPTER 16

# Final Thoughts and Coda

## The Race for an Emissionless Vehicle

Anyone who has read either of my previous two books[1,2] or this book on electric vehicles should realize the difficulty of supplanting the gasoline-powered automobile. This internal combustion vehicle, at its best, is a superb machine for rapidly, safely, inexpensively, and comfortably transporting individuals and goods, regardless of "rain, sleet, snow, or gloom of night."

Despite considerable effort to improve the electric car for more than a century, and particularly in the past 30 years, the battery-only powered car in the mid-1900s at best can occupy only a niche in transportation. Economically, the present electric cars may be likened to small wavelets beating against an almost impregnable, rocky coast—the internal combustion automobile. As you know, the Achilles' heel of the electric car is its limited range. Similarly, the internal combustion motorcar has its own drawback: its emissions, not when there is only one vehicle, but in a city where the number of vehicles may approach a million. As is well known, the present force to mandate electric cars is the reduction of air pollution associated with industrialization. In America, the noxious smokestack has been largely cleared, the waterways have greatly improved, and emissions from individual internal combustion cars have decreased substantially. However, the race is on for an emissionless automobile.

Of the various elements of the electric car, the tires have become a paragon of reliability. Solid-state circuits translate direct-current electricity from a battery into variable-frequency, three-phase power, as shown in the elegant sine waves of Figure 16.1. An induction motor— in its simplicity, low cost, silence of operation, reliability, agelessness, and 95% efficiency—is

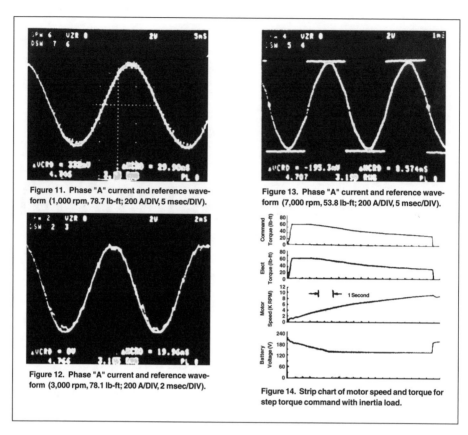

Figure 11. Phase "A" current and reference waveform (1,000 rpm, 78.7 lb-ft; 200 A/DIV, 5 msec/DIV).

Figure 13. Phase "A" current and reference waveform (7,000 rpm, 53.8 lb-ft; 200 A/DIV, 5 msec/DIV).

Figure 12. Phase "A" current and reference waveform (3,000 rpm, 78.1 lb-ft; 200 A/DIV, 2 msec/DIV).

Figure 14. Strip chart of motor speed and torque for step torque command with inertia load.

*Figure 16.1. Three-phase sine wave input to the induction motor. (General Electric)*

at a high plateau of perfection, as shown in Figure 16.2. Likewise, lightweight aluminum cars or cars made from a combination of plastics are being perfected, as illustrated in Figure 16.3. If all this is true, what restrains a mass introduction of the electric car? Already, one top-of-the-line internal combustion automobile contains 36 electric motors performing various functions little dreamed of only 30 years ago.[3] The fact that restrains the mass entrance of the electric car into the market is an absence of an energy source capable of yielding an acceptable vehicle range to satisfy the driver.

The combination of gasoline and the internal combustion engine is simply a world beater and, as a result, has become ubiquitous. Recently, while confined to bed in a fine residential section of a lakeside suburb, I experienced the din from the internal combustion engine. There was no surcease! Road repair equipment, lawn mowers, weed trimmers, leaf blowers, chain saws, outboard motors, mufflerless speed boats, accelerating motorcycles, noisy old cars, garbage trucks, and garden tractors are cacophonous. Little silence exists in any Beantown, and the internal combustion engine is the source of the noise. Contrast that noise with the silence of the electric car or the electric motor.

*Figure 16.2. Cutaway model of an induction motor. (U.S. Department of Energy)*

*Figure 16.3. One of 20 aluminum Ford Mercury Sables for testing that are identical in appearance to normal steel-bodied Sables. The body is 400 lb (181 kg) lighter. Two aluminum Sables passed federal front and rear crash tests. Ford will build 20 more aluminum-intensive vehicles for company and fleet testing. Material cost is $1,350 vs. $675 for a steel shell. (Ford Motor Co.)*

A recent insightful engineering colloquium advanced the low emission of the natural-gas internal combustion engine as the near-term (two decades) answer to a clean automobile capable of long-range trips.[4] The authors of the papers note that the changes required from present production model automobiles are relatively modest. However, what about the infrastructure that is now so abundant for the gasoline automobile? Meanwhile, an almost emissionless energy storage system may be perfected which can, in the vehicle itself, provide a long-range vehicle. The preceding chapters of this book outline promising possible directions, and other solutions may emerge. However, what about the time required for introduction?

Today, a quiet confrontation exists between the government regulators and the automobile manufacturers. The former wish to mandate, and are indeed mandating, emissionless personal vehicles. Each party, I believe, wants to reduce pollution. The crux is the timing. Possibly, reading of my books can, to some degree, reconcile these two divergent positions. If so, then my efforts will have been rewarded.

# Reviewing the Past to Better Understand the Future

The first self-propelled, ground conveyance (other than trains) capable of bearing a person was the early steam carriage. The electric tricycle appeared in 1881, followed by the gasoline-powered tricycle in 1885. The latter two, in attempting to supersede the horse and buggy, required until around 1902, more than 20 years to find the long-term direction—engine in the front and drive wheels in the rear. Already, surely by 1895, the electric and gasoline automobiles, even in their imperfections, were better in some ways than the horse and buggy.

Contrast this with the present. The electric car is chasing a moving target, a continually improved gasoline-powered car. Nevertheless, the race is joined. In *History of the Electric Automobile: Battery-Only Powered* Cars[2] and in chapters of this present book on hybrid electric vehicles, many approaches are shown attempting to find an acceptable complement for the gasoline-powered car. All attempts have fallen short except for niche operations. However, groups are trying. As in 1895, after the eye-popping 704-mile (1133-km) Paris-Bordeaux-Paris race at an average speed of 14.4 mph (23 km/h),[2] workshops around the industrial world started assembling what soon would be called the automobile. Now, in the last third of the twentieth century and almost surely into the first decade of the twenty-first century, the same type of scramble is and will be witnessed. However, the problem today is more difficult. Who knows where the magic, ongoing, electric car design will originate, and what it may be?

For example, one small party, headed by Jonathan Tennyson from the Big Island of Hawaii, is trying. Because Hawaii is an archipelago, a niche may have been found on these small islands for a solar-charged battery electric car. Figure 16.4 shows the Suntera *Sunray* vehicle, its chassis, and its front end. Tennyson told author Michael Hackleman, when visiting the former's workshop,[5]

> Hawaii is ideal for electrics. There's no destination farther than 80 miles.

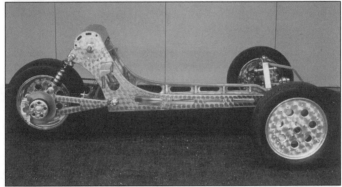

*Figure 16.4. (Top to bottom)  The Suntera* Sunray, *the chassis, and the front end.  (Stephanie Paul. Photos by G. Brad Lewis [top] and Stevi Johnson Paul [center and bottom])*

The *Sunray* bears three wheels (for relaxed regulations), is 8 ft (2.5 m) long, and has a height of 6 ft (2 m); therefore, it can be seen. Its curb weight is 1500 lb (680 kg). Power is from a 12-hp series d-c motor, bearing a 120-volt lead-acid battery system. The drive wheels are belt driven. Figure 16.5 shows the rear drive assembly of the *Sunray*. The *Sunray* comes equipped with solar cells to charge the batteries and thus falls into the class of solar-assist vehicles. Jonathan Tennyson designed the solar-powered *Mana La,* shown in Figure 13.5, for the 1987 Australian World Solar Challenge Race. Figure 16.6 shows designer Tennyson sitting in this beautiful, unique creation.

# Conclusion

Many groups worldwide are experimenting with what is hoped are viable electric motorcars. Will one major design be followed everywhere, as with today's gasoline- and diesel-powered vehicles—the one devised in the early twentieth century? Or is an electric drive system so unique that several basic designs will continue to exist in a market society? The two endemic

*Figure 16.5. Rear drive assembly of the* Sunray. *(Stephanie Paul. Photo by Stevi Johnson Paul)*

*Figure 16.6.  Jonathan Tennyson sitting in the aerodynamically designed* Mana La.
*(Stephanie Paul.  Photo by Mary Van de Ven)*

approaches are:  1) battery-only powered cars, about which I wrote in *History of the Electric Automobile:  Battery-Only Powered Cars*[2], or 2) hybrid electric vehicles, as discussed in this present volume.  Throughout this book, I have referred to the opinions of Dr. Victor Wouk, who is a true pioneer of the modern electric vehicles.  In the future, for a truly successful personal electric automobile, will an engineer emerge such as artist-inventor Samuel F.B. Morse in Baltimore who, on 24 May 1844, first telegraphed to his partner, Alfred Vail in Washington, DC, the prophetic Biblical phrase, "What hath God wrought?"[6]

# References

1.  Wakefield, Ernest Henry, *The Consumer's Electric Car*, Ann Arbor Science Publishers, Ann Arbor, MI, 1977.
2.  Wakefield, Ernest H., *History of the Electric Automobile: Battery-Only Powered Cars*, Society of Automotive Engineers, Warrendale, PA, 1994.
3.  Herbst, Jan F., "Magnets for Electric Motors," General Motors, P.E. Project, Anderson, IN, Lecture, Physics Department, Northwestern University, 1992.
4.  Challenges of Natural Gas Vehicle Technology, SAE TOPTEC, 13–14 April 1992, Society of Automotive Engineers, Warrendale, PA, 1992.
5.  Hackleman, Michael, "Island Electrics," *Home Power,* #45, February/March 1995.
6.  Numbers, Chapter 23, Verse 23.

# The Hughes Inductive Charger System

*Dr. Victor Wouk, the author of this paper, is a well-known electric vehicle consultant. He was asked by General Motors to give an independent appraisal of the charging system developed by Hughes Electric. The following article, reproduced by permission, is based on that appraisal and draws comparisons with the other inductive charging system known as Inductran.[1]*

(In 1994, General Motors placed) 50 *Impact* electric cars with a number of utilities and customers in the United States. In this way, GM hoped to determine how big the market would be if it put such a car into production. The *Impacts* would be fitted with the inductive charging system developed by Hughes (Electric) in Torrance, California.

*Batteries International* first predicted in October 1992 that the Hughes inductive charger worked at very high frequencies based on its high power transfer capabilities. This fact can now be confirmed. Indeed, in evaluating the Hughes Magne-Charge system (as it is called), it is helpful to compare it with the Inductran mains-frequency inductive charging system. This was assessed in the July 1990 issue of *Batteries International.* Magne-Charge was conceived and developed because GM believed (as do other car manufacturers) that, for the electric car to succeed, the charging process must be fast, safe, and hassle-free. It is now possible to see more clearly what GM had in mind.

The Magne-Charge system remains as it was first described two years ago. Instead of a plug and socket, a paddle about the size of a ping-pong bat (Figure A.1) is inserted into a slot in the electric car (Figure A.2). The paddle is connected by a cable to an offboard charging control cabinet. Figure A.3 shows the wall, portable, and floor models. Figure A.4 shows the principle of operation for the Magne-Charge system.

*Figure A.1. The Hughes inductive paddle forms the primary of a transformer working at 80 to 350 kHz. According to preliminary information from Hughes, the same paddle is used for normal charging rates of kilowatts, or opportunity charging at 25 kW, or even opportunity charging at 150 kW. (Hughes Electronics)*

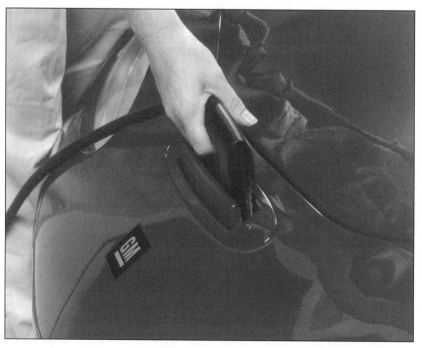

*Figure A.2. Recharge paddle inserted into the car charge port. (Hughes Electronics)*

*Figure A.3. Hughes wall, portable, and floor-mounted charge modules—Magne-Chargers.*
*(Hughes Electronics)*

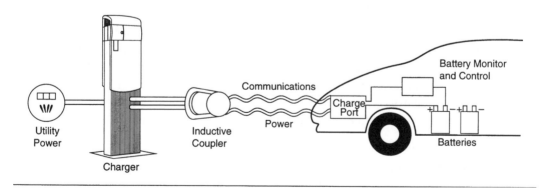

*Figure A.4. Hughes Magne-Charge principle of operation. (Hughes Electronics)*

307

The basic principle of Magne-Charge is the same as the Inductran system. The charging transformer core is in two parts. The part with the primary winding (the paddle) is offboard. The second part with the secondary winding is onboard. Energy flow from the mains power to the rectifiers on the vehicle is through the magnetic flux of the split-core transformer when the primary and secondary are brought into proximity. There the similarity ends.

It is useful to differentiate between the two systems by calling one a proximity device and the other an insertion device. The Inductran is the proximity device. This means that the primary and secondary windings are wound on C-shaped cores which can be separated. To function, the two halves must be as close together as practical; however, they must be only reasonably well aligned. Abutment of the two transformer parts is desirable but not necessary.

In the insertion system, the paddle and its core are effectively a slice taken from a high-frequency transformer. The slice, which consists of part of the core plus the primary winding, must be inserted fully 'home' before the magnetic circuit of the transformer is completed. The lightness of the Magne-Charge paddle means that it can be fixed on the end of handheld flexible cable. This aids the 'mating' operation which is otherwise something of a problem with the Inductran proximity device. However, to keep the copper cables and the paddle light, it does dictate that the a-c current must be high frequency and high voltage.

Inductran is comparatively inflexible in power output because the transformer is designed similarly to a ferro-resonant regulated charger. Thus, it must work around its resonant power level. The batteries fitted to individual trucks can vary in voltage, but their basic power need must be the rating of the Inductran transformer. Inductran is intended for use with materials handling equipment, airport apron equipment, and AGV systems. Magne-Charge is designed basically for electric cars. Therefore, wide flexibility is required in voltage output and current charging regimes. This flexibility makes Magne-Charge inherently more complex and expensive than Inductran. However, industrial applications possibly can be found for Magne-Charge, as will be discussed later.

# How Magne-Charge Works

The offboard control cabinet of Magne-Charge contains complex circuitry that converts 60 Hz mains power into high-voltage, low-current a-c at frequencies of 80 to 350 kHz. The control cabinet controls the charging regime in a manner demanded by the onboard circuits that are responsive to the voltage of the battery, its state of charge, etc. The charging needs are signaled to the control circuits via a low-power radio-frequency (RF) signal generated on the electric vehicle (EV) and linked to the control circuits via a pickup coil in the paddle.

As in the Inductran system, no current-carrying circuits are exposed. No arcing occurs if the paddle is withdrawn from the charge port under load. Apprehensions about the EV being ungrounded during charging are unfounded, according to Hughes. Its analysis, supported by experience with the Inductran system on materials handling equipment, indicates that no hazard exists with ungrounded trucks. On the contrary, it is claimed that when the floor is wet, the vehicle presents a decreased hazard. Because no return path is found from the vehicle to the mains-frequency source, only transformer leakage-current can flow if part of the secondary of the high-frequency transformer becomes grounded to the vehicle metalwork. Hughes claims that this current is below physiological reaction levels, as it is with Inductran.

Safety has been a major concern in the design of Magne-Charge because the paddle must be handled and inserted manually. The main safety feature is that the system is not allowed to energize unless the paddle is fully inserted into the charge port and is properly seated. If the paddle is withdrawn, it is automatically de-energized. Other construction and safety features are incorporated into the Magne-Charge system to make it resistant to misuse and vandalism. A promotional video shows an energized paddle carrying full charging current in a goldfish tank. The fish swim around the tank as usual, regardless of whether or not current is flowing.

High frequency, as stated previously, was chosen because of the need to reduce weight. *Impact's* battery stores 16.7 kWh of energy. To complete a charge in two to three hours means fitting a charger of 6-kW capacity which would weigh at least 50 kg (110 lb) for a complete 60-Hz unit with transformer. Inductran effectively halves the weight carried onboard to the vehicle. Thus, the onboard portion of a 3-kW Inductran system weighs approximately 20 kg (44 lb). By comparison, the onboard part of a Magne-Charge unit of twice the power (6 kW) weighs only 9 kg (20 lb), and most of this weight is taken up with the rectifiers and cables rather than the transformer. Hughes appears to say that this same onboard portion could handle power levels up to 25 kW for fast charging.

Hughes has developed three versions of the Magne-Charge cabinets that transmit the power. These are the units that accept mains power and convert it into high-frequency power for the paddle. Table A.1 shows the electrical package configurations of three initial models of Magne-Charge.

1. The Standard Charge Module (SCM) is rated at 6.2-kW output and is intended for charging a single vehicle such as *Impact*. Thus, in the GM demonstration program, one of these chargers will be located at the home base of each vehicle.

2. The Opportunity Charge Module (OCM) is being developed for interim or 'opportunity' charging in urban areas. The first model for which data are given is rated at 25 kW, but later units will be rated up to 150 kW.

3. The Convenience Charge Module (CCM) is a small transportable version of the SCM. It is rated at 1.2 kW and weighs 7.5 kg (16.5 lb). This can be plugged into a standard outlet (120 V, 20 A, 60 Hz in the United States) and delivers its output into a paddle in the same way as its elder brothers.

Likewise, Table A.2 summarizes the other features of this charging system.

## TABLE A.1
## POWER AND PACKAGE—MAGNE-CHARGE

### Power Input

| | Standard Charger | Portable Charger |
|---|---|---|
| Volts | 165 to 260 a-c | 90 to 135 a-c |
| Amperes | 0.1 to 30 maximum | 0.1 to 12 maximum |
| Hertz | 50/60 | 50/60 |

### Power Output

| | Standard Charger | Portable Charger |
|---|---|---|
| Volts | 100 to 405 d-c | 100 to 405 d-c |
| Amperes | 2 to 18 maximum | 0.5 to 3.5 maximum |
| Kilowatts | 0.2 to 6.1 | 0.2 to 1.2 |

### Package Configurations

| | Standard Wall-Mounted Charger | Standard Floor-Mounted Charger | Portable Charger |
|---|---|---|---|
| Height, mm (in.) | 825 (32.5) | 1442 (57) | 330 (13.0) |
| Width (mm) (in.) | 298 (11.8) | 354 (14) | 216 (8.5) |
| Depth, mm (in.) | 286 (11.3) | 304 (12) | 114 (4.5) |
| Weight, kg (lb) | 27.7 (61) | 39 (86) | 7.5 (16.5) |
| Environmental | Indoor | All weather | All weather |

Source: Hughes Electronics.

## Electrical Performance

Figure A.5 shows the overall electrical efficiency, power factor, and total harmonic distortion of the SCM, OCM, and CCM units taken together with the matching charge port on the vehicle. Hughes says the information is preliminary and has not been measured on production hardware. Hughes also makes the observation that the efficiency, power factors, and levels of electromagnetic interference (EMI) are 'acceptable.'

## Availability and Price

Production units of the SCM were due to become available only after April 1994, in time for delivery to U.S. electricity utilities who are taking *Impact* vehicles. The recommended retail price of the 6-kW wall-mounted indoor unit will be $4,250, or $500 more for a post-mounted weatherproof unit. The receiver fitted to the car (the charge port) is $95. Sale prices are projected by Hughes to be less that $1,000 for the control cabinet and $200 for the charge port when made in 'Detroit-produced quantities.'[a,2]

[a] Southern California Edison Company's Diane Wittenberg, vice president for electric vehicles, said her company expects to provide approximately 8,000 charging stations in the next three years. See Ref. 2 for more information.

## Performance (Charger With Charge Port)*

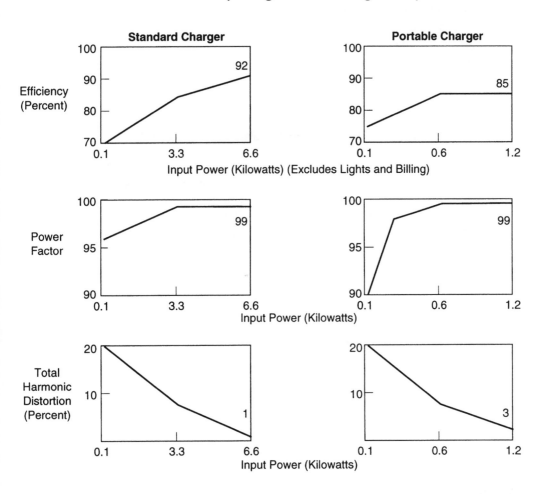

*Information is based on engineering systems.

*Figure A.5. Overall efficiency, power factors, and total harmonic distortion (preliminary data).
(Hughes Electronics)*

The first 25-kW OCM unit became available in September 1994, carrying a price of $71,430. It will clearly be capable of servicing a large number of vehicles (although not simultaneously), depending on whether they are given at partial or full recharge. It is tempting to ask whether the high-frequency paddle system could prove attractive to industrial users. At present, Hughes has not assessed the market, but several electricity utilities have taken on distribution rights which would enable them to sell to industrial users. The data quoted in this article will allow companies to decide whether they could be interested in the system.

Perhaps it is helpful to conclude with another mention of Inductran. Compared with Magne-Charge in small quantities, Inductran is a relatively low-cost system but needs further improvements in simplifying alignment for electric car application. Magne-Charge overcomes the alignment problem, but the promised cost reductions must occur if it is to find widespread use compared with existing technology.

**TABLE A.2**
**OTHER FEATURES OF THE MAGNE-CHARGE SYSTEM**

### Vehicle to Charger

| Interface | Functions/Features |
|---|---|
| Bidirectional data bus | Charging control |
| Close proximity RF (915 MHz) | Delayed charging |
| 10.4 kb/s data rate | Vehicle preconditioning |
| SAE J1850 protocol | |

### Charger to Home/Utility/Billing

| Interface | Functions/Features |
|---|---|
| Microprocessor has RS232 port available | Utility load management |
| 10.4 kb/s data rate | Automatic billing |
| Protocol in-work with utilities, home automation, and banks | Vehicle monitoring |
| | Remote monitoring |

Source: Hughes Electronics.

# References

1.  Wouk, Victor, "The Hughes Inductive Charger System," *Batteries International,* 19 April 1994, pp. 52–57.
2.  *The New York Times*, 19 June 1996, p. C3.

# Ultracapacitors for Electric Automobiles

A capacitor may be defined as a device that consists essentially of two conductors (such as parallel metal plates) insulated from each other by a dielectric. This type of assembly may introduce capacitance into a circuit, can store electric energy, can block the flow of direct current, and can permit the flow of alternating current to a degree dependent on the capacitance of the capacitor and the current frequency. Its unit is the Farad, named after England's Michael Faraday (1781–1867). A Farad (F) is defined as the capacitance of a capacitor between the plates of which there appears a potential difference of one volt when it is charged by a quantity of electricity equal to one coulomb. Inasmuch as a Farad is comparatively large, the microFarad ($\mu$F, where $\mu$F = $10^{-6}$ F) and the picoFarad (pF, where pF = $10^{-12}$ F) are commonly used. *History of the Electric Automobile: Battery-Only Powered Cars*[1] contains a glossary of terms useful in electric cars.

The concept of capacitance in an electric circuit was first illuminated by Ewald George von Kliest and Peter von Musschenbroek who independently discovered (1745–1746) electricity could be stored in a glass bottle if containing copper wire or mercury. By 1748, the glass bottle was coated on the outside and inside with separate metal foils, with the inner connected to a conducting rod and terminated in a conducting sphere. The jar usually was charged with an electrostatic machine. On discharge, small animals could be killed. From this experiment, Grolatt created at the University of Leyden what became known as the Leyden jar, an object which today is recognized as an electrical capacitor.[2]

Beginning in the early 1990s, serious consideration was accorded the use of the electric capacitor as an energy storage device, particularly for transients, for electric vehicles. The reasoning was as follows here.[3] In addition, see Ref. 4 for a splendid field mapping and computational model for the electric field of two parallel conducting plates separated by a dielectric.

High energy and power density ultracapacitors can be used to accept regenerative braking power and provide accelerating power, allowing for more uniform or level loads to be handled by traction batteries. For electric vehicles, power to sustain level loads is less than one third of peak power requirements. Therefore, ultracapacitors, in combination with batteries, can increase electric vehicle range by as much as 50 percent for some types of batteries, and more than double the cycle life. The size (kilowatt-hours stored) and thus the cost of the battery system needed to meet vehicle range requirements could potentially be reduced. Ultracapacitors are being considered for regenerative braking energy storage and accelerating power in conjunction with hybrid propulsion systems.

The ultracapacitor development program for electric and hybrid application is outlined in Table B.1. Near-term and advanced program goals have been defined for specific energy (5 and 15 watt-hours per kilogram) and specific power (500 and 1600 watts per kilogram). The Idaho National Engineering Laboratory is responsible for technical program management, technology assessment, and testing. Several carbon-based and non-carbon-based technologies are being developed for capacitors with specific energies exceeding the near-term goal of 5 watt-hours per kilogram. Several devices were tested during fiscal year 1995.

Non-aqueous (organic electrolyte 3 volts per cell), carbon-based 24-volt bipolar stacks have been fabricated (laboratory fixtures and packaged) by Maxwell Laboratories, Inc. and tested at the Idaho National Engineering Laboratory. The specific energy goal was met easily when calculated using the weight of active materials only. Maxwell's first embodiment of a packaged 24-volt device suffered some performance loss and fell slightly below the goals at approximately 4.5 watt-hours per kilogram and 500 watts per kilogram. Two-cell (6-volt) and three-cell (9-volt) devices with lighter metal foil packaging also were delivered. These two devices performed slightly better than the 24-volt stacks with specific energies meeting or exceeding goals at 5 watt-hours per kilogram at 1000 watts per kilogram. The 24-volt stacks are approximately 2 Farads. All cells presented are approximately 20 square centimeters at 0.8 Farads per square centimeter. Increasing the carbon loading in the electrodes is expected to increase specific energies to these cells to 10 watt-hours per kilogram.

Lawrence Livermore National Laboratory delivered an 8-cell 6.5-volt Aerocapacitor (carbon/aqueous) packaged device, rated at 0.19 Farads, for testing at the Idaho National Engineering Laboratory. Based on the weight of active materials (estimated), specific energies are less than 1 watt-hour per kilogram at specific powers of 200 watts per kilogram or higher. This device represents a new thin cell design which shows promise. Work is continuing on adapting this system to high voltage (3 volts per cell) organic electrolytes so that these electrode materials can be evaluated against other carbon/organic systems under development.

Federal Fabrics delivered a 1-volt cell, 120-Farad device containing high-density carbon electrodes (1 gram per cubic centimeter) and an aqueous acid electrolyte. The power capacity of the cells was limited by its high resistance of more than 2 ohms per square centimeter. Specific

**TABLE B.1**
**STATUS OF ULTRACAPACITOR TECHNOLOGIES**

| Technology | Developer | Early Deliverables | | | Expected Deliverables | | | Projected Energy Density (W-h/kg) |
|---|---|---|---|---|---|---|---|---|
| | | Date | Description | W-h/kg | Date | Description | W-h/kg | |
| Composite Carbon Fiber | Maxwell/Auburn | 7/93 | Carbon/nickel/aqueous cells, 20 cm$^2$ | 1.5–2.0 | 9/96 | Carbon/alum/organic 200 cm$^2$, 48-volt cells | 5–7 | 10–12 |
| Aerogel Carbon | Lawrence Livermore National Laboratory | 10/93 | Aerogel carbon, aqueous cells, 80 cm$^2$ | 1–2 | 1/96 | Aerogel carbon, organic cells, 80 cm$^2$ | 5–7 | 10–12 |
| Mixed-Oxide (Ceramic) | Pinnacle Research Institute | 11/94 | Mixed-oxide, aqueous 28-volt stack, 80 cm$^2$ | 1 | | | | |
| Foamed Carbon Particulate with Binder | Sandia National Laboratories | 11/93 | Foamed carbon with binder, aqueous, 1-volt cells | 2–3 | 1/96 | Carbon/alum/organic 200 cm$^2$ cells | 5–8 | 8–10 |
| Doped Polymer on Carbon Paper | Los Alamos National Laboratory | 9/94 | Doped polymer, organic (Type I) 1-volt, cell | 2 (5 based on active polymer material) | | | | |
| Z-axis Carbon | Federal Fabrics | 11/94 | Z-axis carbon, aqueous, 8 cm$^2$, 1-volt cells | 1 | 12/95 | Z-axis carbon, aqueous, 20 cm$^2$, 20-volt stack | 5–7 | 10 |
| | | | | | 9/96 | Z-axis carbon, organic, 20 cm$^2$, 50-volt bipolar | 8–10 | 15 |
| Nanostructure Multilayer | Lawrence Livermore National Laboratory | 9/94 | ZrO$_2$/Al$_2$O$_3$, 1200-volt 813 pF device | 1 | 6/96 | Multilayer, 1 μ, 400-volt, 50 cm$^2$, very low loss (<0.1%) | 2–3 | >3 |

energies were calculated at less than 1 watt-hour per kilogram using current drains of less than 0.5 amps. The high-density electrodes are the highest capacity yet delivered. Technical challenges remain, including achieving electrolyte penetration into high-density carbon electrodes.

The Pinnacle Research Institute fabricated an 80-square-centimeters, 0.8-Farads, 28-volt bipolar stack device containing 28 1-volt cells in an aqueous acid electrolyte and delivered it to Idaho National Engineering Laboratory for testing. The electrodes of the bipolar stack are composed of porous ruthenium and tantalum oxide films. Although not a carbon-based chemistry, this device is presented for comparison because it represents the most advanced packaging of ultracapacitor technology in high voltage bipolar stacks. At the 500-watts-per-kilogram constant power discharge level, the devices were capable of delivering only approximately 0.4 watt-hours per kilogram.

Considerable progress in the development of nanostructure multilayer technology was achieved by Lawrence Livermore National Laboratory. This solid-state technology utilizes deposition manufacturing processes to surpass the performance of other devices operating under the traditional dielectric principle. A number of dielectrics, conductors, and substrate materials were investigated, and automated fabrication processes were refined. In addition to a number of prototypes developed, two large-area (23-square-centimeters) two-layer capacitors were produced that demonstrate the viability of the technology.

An *Electric Vehicle Capacitor Test Procedures Manual* (DOD/ID-10491) was completed by the Idaho National Engineering Laboratory and JME, Incorporated to provide consistent test methods for evaluating the performance of electrochemical capacitors intended for use in electric and hybrid vehicle drivelines. Consistent testing guidelines were needed to compare one technical approach or capacitor product configuration with another. The standardized test methods and reporting can identify critical performance deficiencies, thereby providing direction for subsequent developments.

# References

1. Wakefield, Ernest H., *History of the Electric Automobile: Battery-Only Powered Cars*, Society of Automotive Engineers, Warrendale, PA, 1994.
2. Considine, Douglas M., ed., *Van Nostrand's Scientific Encyclopedia,* 8th ed., Van Nostrand Reinhold, New York, 1995.
3. "Electric and Hybrid Vehicles Program," 19th Annual Report to Congress for Fiscal Year 1995, U.S. Department of Energy Assistant Secretary, Energy Efficiency and Renewable Energy Office of Transportation Technologies, June 1996.
4. Atwood, Stephen S., *Electric and Magnetic Fields*, 3rd ed., J. Wiley & Sons, New York, 1949, pp. 74–75.

# Index

Abbreviations are used after the page number to indicate figure (*f*), footnote (*n*), or tabular material (*t*).

# About the Author

For almost 40 years, Dr. Ernest Henry Wakefield has been involved with work on electric vehicles in some way, with diverse experience. In his teenage years, he authored a newspaper column in addition to building, repairing, and owning eleven boats: skiffs, power, sail, and auxiliary. He also served as an oiler on the *S.S. Helen* on the North Atlantic. His education includes a B.A., an M.S. (Fine Arts), and a Ph.D. (Electrical Engineering) from the University of Michigan. He was employed at General Electric Company and Westinghouse Corporation and taught electrical engineering at the University of Tennessee. Dr. Wakefield volunteered into the U.S. Army, with military training at Camp Crowder, Missouri, and did stints at Massachusetts Institute of Technology in radar and the University of Chicago in physics. He was assigned to the Physics Division of The Manhattan Project (atomic bomb), assisting Enrico Fermi, who was the so-called "father" of the nuclear reactor at the University of Chicago. In this expanding world, Dr. Wakefield continues to write.

After World War II, Dr. Wakefield founded, owned, and operated a nuclear instrument company with four factories in Skokie, Illinois, and Berkeley, California, manufacturing for both the domestic and export markets. Later, he independently discovered the concept of variable frequency and pulse-width modulation for changing direct-current to variable frequency three-phase alternating-current, a technique now being applied to electric vehicles, locomotives, and fixed industrialized machines worldwide. To employ this concept, he designed and built both d-c and a-c powered electric cars.

During this period, Dr. Wakefield continued his writings on transportation, authoring *The Consumer's Electric Car* (1977), *History of the Electric Automobile: Battery-Only Powered Cars* (1994), and the present book on the history of hybrid electric vehicles. He has traveled in 59 and lectured in more than 20 foreign countries, and participated for more than

40 years as a visitor in both the African and Physics programs at Northwestern University in Evanston, Illinois. He currently resides near his daughter in Silver Spring, Maryland. His other writings include works about the nuclear field, entrepreneurship, the American Civil War, economic papers on Third World nations, and some fiction.